3G Wireless with WiMAX and Wi-Fi: 802.16 and 802.11

3G Wireless with WiMAX and Wi-Fi: 802.16 and 802.11

Clint Smith, P.E.

John Meyer

McGraw-Hill

New York Chicago San Francisco Lisbon London Madrid
Mexico City Milan New Delhi San Juan Seoul
Singapore Sydney Toronto

The McGraw·Hill Companies

Library of Congress Cataloging-in-Publication Data

Smith, Clint,
    3G Wireless with WiMAX and Wi-Fi: 802.16 and 802.11/Clint Smith,
    John Meyer.
        p.   cm.
    ISBN 0-07-144082-8
    1. Mobile communication systems.   2. Wireless communication
    systems.   3. Global system for mobile communications.
    I. Meyer, John. II. Title.

    TK6570.M6S595   2004
    621.3845'6—dc22

                                                    2004054659

1 2 3 4 5 6 7 8 9 0 DOC/DOC 0 1 0 9 8 7 6 5 4

ISBN 0-07-144082-8

*The sponsoring editor for this book was Steve Chapman and the
production supervisor was Sherri Souffrance. It was set in Century
Schoolbook by International Typesetting and Composition. The art
director for the cover was Margaret Webster-Shapiro.*

Printed and bound by RR Donnelley.

This book was printed on recycled, acid-free paper containing
a minimum of 50% recycled, de-inked fiber.

McGraw-Hill books are available at special quantity discounts to use as
premiums and sales promotions, or for use in corporate training programs. For more information, please write to the Director of Special
Sales, McGraw-Hill Professional, Two Penn Plaza, New York, NY 10121-
2298. Or contact your local bookstore.

To Mary and our two children who endure my quest to document this exciting industry. And to all my many colleagues who constantly help in providing needed guidance and direction.

Clint Smith, P.E.

To Cindy for all the support she provides.

John Meyer

Contents

Preface

The convergence of wireless mobility and 802.11 is happening; in particular, in the area of 802.11b,g,a (commonly known as Wi-Fi). There are announcements of new Wi-Fi hot spots being made available for use every day. Wi-Fi, although available for many years, is currently in its infancy with respect to wireless mobility. With the proliferation of hot spots and the global reach of 3G mobile wireless services, the possibility of ubiquitous mobility for data transport is a reality for Cellular and PCS operators who blend Wi-Fi systems into their service offerings.

This simple concept requires a fundamental shift in the thinking and operating philosophy for a wireless operator. It involves how to glean revenue from this new service. It may be problematic to charge the customer for using license-exempt spectrum that others are using for free. A more palatable method may be to make this a part of their high-speed data offering, thus attracting more data users. And of course there is a great deal of value in reducing customer churn.

It is too early to determine whether or not mobile wireless companies can achieve an increase in the average revenue per user (ARPU) due to Wi-Fi services alone. However, 802.11 has many unique advantages and is becoming a standard platform for computing devices in the same fashion as Network Interface Cards (NICs) were several years ago. Ethernet ports, once an option are now standard on most PCs. Wi-Fi capability will soon be just as commonplace. With every new Wi-Fi hot spot being introduced into the market, the pressure builds for Cellular and PCS providers to offer Wi-Fi integration in their existing 2.5/3G Network. If not, there is the possibility of losing data revenue and possibly customers to Wi-Fi operators.

The advent of 802.16 (sometimes referred to as WiMAX) also has entered the mix. In addition to providing a standard for wide area wireless networks it offers the potential to provide the necessary backhaul for 802.11. This could substantially reduce the backhaul costs associated with supporting Wi-Fi hot spots. A combination of Wi-Fi/WiMAX ubiquity and wireless VoIP could pose a threat to the 3G mobile operators.

WiMAX/802.16 also addresses a fundamental stumbling block for a number of spectrum license holders. In particular, Local Multipoint Distribution Systems/Metropolitan Multipoint Distribution Systems (LMDS/MMDS) operators will be affected. Due to the lack of standardization for use of these bands, the selection of infrastructure solutions was difficult. More importantly, there were no economies of scale available for customer equipment. If LMDS/MMDS can approach the pricing of Wi-Fi equipment there will be a resurgence of interest in these areas.

This book will cover many of the aspects that wireless mobility operators need to consider in the design and integration of 802.11 into their system. Background on 3G mobile wireless systems will also be provided. The issues of mobility are very similar in Wi-Fi and mobile wireless. The integration of these two systems seems inevitable.

Wi-Fi integration can be as simple as providing a bandwidth pipe to a Wi-Fi hot spot. Alternatively, the 802.11 solution may be an enterprise solution for the mobile workforce, involving the seamless blending of Wi-Fi and the 3G network.

But no mater what the situation, security and integrity will be an essential part of the solution. While Wi-Fi has rudimentary security built-in, additional measures will be required for the corporate network environment. In addition to enhanced security methods there will be an increased need for flexible Virtual Private Network systems geared toward mobility.

The potential impact of 802.16 will also be discussed. Applications of 802.16 as a part of the backhaul solution for Wi-Fi will be examined. 802.16 (WiMAX) will also have an impact on other operators including LMDS and MMDS networks.

Lastly, 802.20 too will be considered. Issues related to ensuring that platforms deployed by mobile wireless operators will not become obsolete, further complicating the legacy system problems the wireless mobility operators have today, will be dealt with.

The book will introduce the reader to the various aspects of 802.11 ranging from technology discussions to the practical guidelines. The use of 802.16 and 802.20 will also be covered. This will hopefully assist in decision making for Wi-Fi deployment and integration going forward.

In short, we sincerely hope that this book will help you in making the correct decisions.

John Meyer
Clint Smith, P.E.

ABOUT THE AUTHORS

CLINT SMITH, P.E. is the Vice President of ASI, where he is currently involved in the design and deployment of mobile data networks with GSM/GPRS/EDGE and CDMA2000 from both a core network and RF aspect. He has extensive design and operational experience in both mobile and fixed networks. He is also the author of *Practical Cellular and PCS Design, LMDS, Wireless Telecom FAQ* and a co-author of *Cellular System Design and Optimization, 3G Wireless Networks* and *Wireless Network Performance Handbook.* Clint holds a Masters in Business Administration from Fairleigh Dickinson University and a Bachelor of Engineering degree from Stevens Institute of Technology. He is also a registered professional engineer.

JOHN MEYER is Vice President of Engineering Services at Velocitel, Inc. An expert in both network and RF engineering, he has over 25 years' experience in the communications industry, including 10 years with the Motorola Cellular Infrastructure Group. He is actively involved with the deployment and integration of 2.5G and 3G mobile data networks with an emphasis on both the financial and technical components of offering wireless mobile data. John has a BS in Computer Science and a MBA. He lives in Oakbrook, Illinois.

3G Wireless with WiMAX and Wi-Fi: 802.16 and 802.11

1

Introduction

The mobile wireless industry is poised for greatness. The industry, over the last few years, has undergone and continues to undergo a tremendous amount of change. The change is brought about through the introduction of a never-ending stream of technologies all designed to provide unique services that customers will purchase. The plethora of technology innovations are at the infrastructure and handset level, besides all the applications that are emerging. The fundamental question that everyone is or should be asking is "Where will this all go?"

At the heart of all the technology platforms and handsets introduced is *access*—being able to access both voice and data services regardless of where the end user is physically located. Until the last few years mobility had always been an adjunct service to the landline services in developed countries. Now with Wireless Number Portability (WNP) the mobile wireless industry continues striving to augment or even replace the wired local loop. This effort has fostered numerous radio technologies that operate over a vast range of the spectrum—from 400 MHz to 40 GHz.

The initial concepts of wireless access involved delivering voice services—whether they were analog or digitized voice—using a host of modulation techniques. The primary focus was on the deployment of radio base stations and then the development of adjunct services from which customer retention and enhanced revenues could be exploited. But as the internet and other bandwidth-hungry services and products have become more prolific in society, data followed by voice services are what is envisioned.

This chapter is meant to provide a brief sampling of the vast array of wireless mobility concepts. The mobility concepts should essentially be understood as convergence between voice and data. Legacy systems used for deploying wireless mobility systems involving 1G, 2G, 2.5G, and 3G will both enhance and hamper the wireless and wireline convergence.

The use of analog mobile platforms associated with 1G, while still prevalent in the world will not be discussed. They are not included in the convergence because of their fundamental lack of packet data capability.

1.1 Cellular Concept

Cellular communication is one of the most prolific mobile voice communication platforms that have been deployed. The concept of Cellular Radio was initially developed by AT&T at their Bell Laboratories to provide additional radio capacity for a geographic customer service area. The initial mobile systems from which cellular evolved were called *Mobile Telephone Systems* (MTS). Later, improvements to these systems were made and the systems were referred to as *Improved Mobile Telephone Systems* (IMTS). One of the main problems with these systems is that a mobile call could not be transferred from one radio station to another without loss of communication. This problem was resolved by implementing the concept of reusing the allocated frequencies of the system. Reusing the frequencies in cellular systems enables a market to offer higher radio traffic capacity. The increased radio traffic allows more users in a geographic service area than was possible with the MTS or IMTS systems.

Cellular Radio was a logical progression in the quest to provide additional radio capacity for a geographic area. The cellular system as it is known today has its primary roots in the MTS and the IMTS. Both are similar to Cellular with the exception that no handoff takes place with these networks.

Cellular systems operate on the principle of frequency reuse. Frequency reuse in a cellular market provides a cellular operator the ability to offer higher radio traffic capacity. The higher radio traffic capacity enables many more users in a geographic area to use radio communication than was possible with an MTS or IMTS system.

A brief high-level system configuration for a cellular system is shown below in Fig. 1.1.

There are numerous types of Cellular systems that are used both in the United States and elsewhere. All the systems are similar in network layout in that they have base stations connected to a *Mobile Switching Center* (MSC) which in turn connects to the PSTN or PTT.

The cellular concept is employed in many different forms. Typically, when referencing cellular communication it is first applied to either the *Advanced Mobile Phone System* or *Total-Access Communications System* (AMPS or TACS) technology. With AMPS operating in the 800 MHz band the frequency range is 821 to 849 MHz for base station

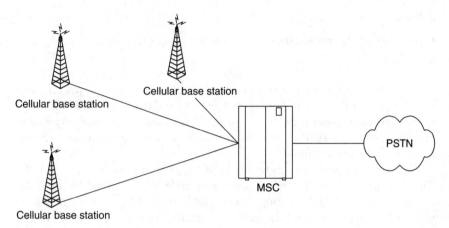

Figure 1.1   General cellular system configuration.

receive and 869 to 894 MHz for the base station transmit. For TACS, the frequency range is 890 to 915 MHz for base receive and 935 to 960 MHz for base station transmit.

Many other technologies also fall within the guise of cellular communication and they involve the PCS bands, both domestic (U.S.) and the international bands. In addition, the same concept is applied to several technology platforms that are currently used in the *Specialized Mobile Radio* (SMR) band (IS-136 and *Integrated Dispatch Enhanced Network,* iDEN). However, cellular communication is really referenced to both the AMPS and TACS bands but is sometimes interchanged with the PCS and SMR bands because of the similarities.

Personal Communication Service (PCS) was described at the time the frequency bands were made available as the next generation of wireless communication. PCS, by default, has similarities and differences with its counterparts in the cellular band. The similarities between PCS and Cellular lie in the mobility of the user of the service. The differences between PCS and cellular are in the applications and spectrum available to PCS operators for providing services to their subscribers.

Digital or digital modulation is now prevalent throughout the entire wireless industry. Digital communication refers to any communication that utilizes a modulation format that relies on sending the information in any type of data format. More specifically, digital communication is where the sending location digitizes the voice communication and then modulates it. At the receiver, exactly the opposite is done.

Mobile communication, whether it is called Cellular or PCS is the form of wireless communication that enables several key concepts to be employed:

- Frequency reuse
- Mobility of the subscriber (i.e., location independence)
- Handoffs

There are numerous wireless mobile technology platforms in existence today, and more are likely to come in the future. Some of the platforms are born from standards and others from proprietary systems. However, each particular wireless technology has its own unique advantages and disadvantages.

Presently the wireless industry is in transition between 2G and 3G. There is an interim set of technology platforms referred to as 2.5G. What exactly is the definition of 2G, 2.5G, and 3G is a constant source of debate. However, the following is a simple definition for each of the generations.

- First Generation (1G) includes all the analog technologies, primarily AMPS and TACS.
- Second Generation (2G) includes GSM, IS-136 and IS-95, and iDEN.
- Two and one-half Generation (2.5G) includes GSM/GPRS/EDGE, CDMA2000 1xRTT, and 1xDO, iDEN, and WiDEN
- Third Generation (3G) include WCDMA and CDMA2000 (EVDO, EVDV)

1.2 Generic Wireless System Configuration

Wireless systems are commonly called Cellular. This simply means that the network is divided into a number of cells as shown in Fig. 1.2. Within each cell is a base station, which contains the radio transmission and reception equipment. It is the base station that provides the radio communication with those mobile phones that happen to be within the cell, where the cell is simply the geographical coverage area of a given base station. The coverage area of a given cell is dependent upon a number of factors such as the transmit power of the base station, the transmit power of the mobile, the height of the base station antennas, and the topology of the landscape. The coverage of a cell can range from as little as about one hundred yards to tens of miles.

Specific radio frequencies are allocated within each cell in a manner that depends on the technology in question. In most systems, a number of individual frequencies are allocated to a given cell and those same frequencies are reused in other cells that are sufficiently far away to avoid interference. With CDMA, however, the same frequency can be reused in every cell.

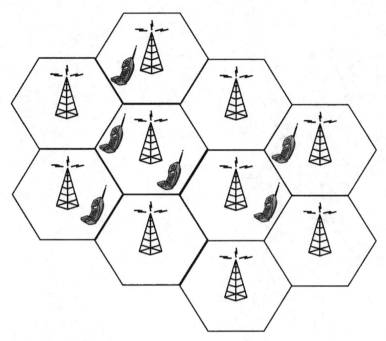

Figure 1.2   Cellular system.

While the scheme shown in Fig. 1.2 is certainly feasible and is some-times implemented, it is common that cells are sectored as shown in Fig. 1.3. In this approach the base station equipment for a number of cells is colocated at the edge of these cells and directional antennas are used to provide coverage over the area of each cell (as opposed to omni-directional antennas where the base station is located at the center of a cell). Sectored arrangements with up to six sectors are known, but the most common configuration is three-sectors per base station in urban areas with two-sectors per base station along highways.

Of course, it is necessary that the base stations be connected to a switching network and for that network to be connected to other net-works such as the *Public Switched Telephone Network* (PSTN), so that calls can be made to and from mobile subscribers. Furthermore, it is nec-essary for information about the mobile subscribers to be stored in a par-ticular place in the network. Given that different subscribers may have different services and features, the network must know what services and features apply to each subscriber in order to handle calls appro-priately. For example, a given subscriber may be prohibited from making international calls. Should the subscriber attempt to make an interna-tional call, the network must disallow that call based on the subscriber's service profile.

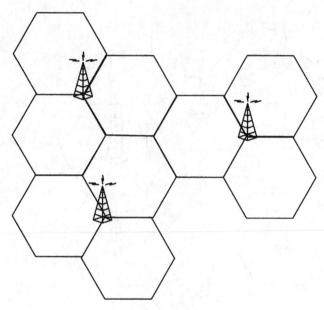

Three-sector configuration

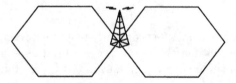

Two-sector configuration

Figure 1.3 Typical sectored cell sites.

Figure 1.4 shows a typical (although very basic) mobile communication network. A number of base stations are connected to a *Base Station Controller* (BSC). The BSC contains logic to control each of the base stations. Among other tasks, the BSC manages the handoff of calls from one base station to another as a subscriber moves from cell to cell. Note that in certain implementations the BSC may physically and logically combine with the MSC.

Connected to the BSC is the *Mobile Switching Center* (MSC). The MSC, also known in some circles as the *Mobile Telephone Switching Office* (MTSO), is the switch that manages the set-up and tear-down of calls to and from mobile subscribers. The MSC contains many of the features and functions found in a standard PSTN switch. It also contains, however, a number of functions that are specific to mobile communication.

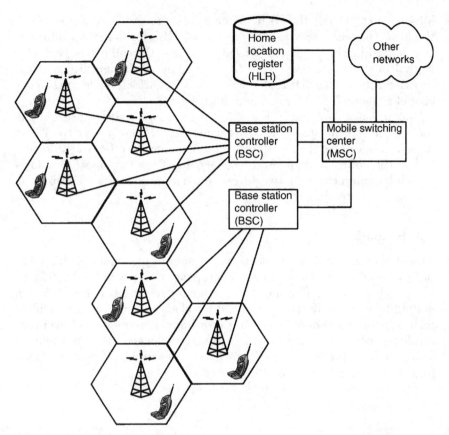

Figure 1.4   Basic network architecture.

For example, the BSC functionality may be contained with the MSC in certain systems, particularly in 1G systems. Even if the BSC functionality is not contained within the MSC, the MSC must still interface with a number of BSCs over an interface that is not found in other types of networks. Furthermore, the MSC must contain logic of its own to deal with the fact that the subscribers are mobile. Part of this logic involves an interface to one or more *Home Location Registers* (HLRs), where subscriber-specific data are held.

The HLR contains subscription information related to a number of subscribers. It is effectively a subscriber database and is usually depicted in diagrams as a database. The HLR does, however, do more that just hold subscriber data; it also plays a critical role in mobility management, i.e., the tracking of a subscriber as he or she moves around the network. In particular, as a subscriber moves from one MSC to another, each

MSC in turn notifies the HLR. When a call is received from the PSTN, the MSC that receives the call queries the HLR for the latest information regarding the subscriber's location so that the call can be correctly routed to the subscriber. Note that, in some implementations, HLR functionality is incorporated within the MSC, which leads to the concept of a "home MSC" for a given subscriber.

The network depicted in Fig. 1.4 can be considered to represent the bare minimum needed to provide a mobile telephony service. These days, a range of services with different features are offered in addition to just the ability to make and receive calls. Therefore, most of today's mobile communication networks are much more sophisticated than the network depicted in Fig. 1.4.

1.3 Handoffs

The handoff concept is one of the fundamental principles of this technology. Handoffs enable cellular to operate at lower power levels and provide high capacity. The handoff scenario presented in Fig. 1.5 is a simplified process. There are a multitude of algorithms that are invoked for the generation and processing of a handoff request and eventual handoff order. The individual algorithms are dependent upon the individual vendor for the network infrastructure and the software loads and specific radio access platform used.

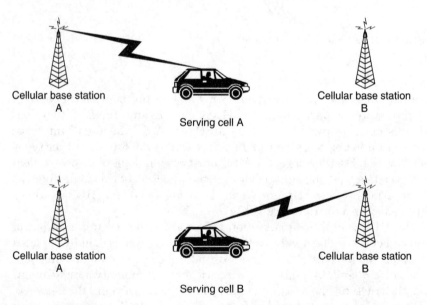

Figure 1.5 Handoff.

Handing off from cell to cell is fundamentally the process of transferring the mobile unit that has a call in progress on a particular voice channel to another voice channel, all without interrupting the call. Handoffs can occur between adjacent cells or sectors of the same cell site. The actual need for a handoff is determined by the actual quality of the RF signal received from the mobile into the cell site.

As the mobile transverses the cellular network it is handed off from one cell site to another cell site ensuring a quality call is maintained for the duration of the conversation.

1.4 Typical Central Office (CO)

A central office is anything but typical. While a CO typically delivers voice service and provides the local loop aspect for telephony the particular function and services the CO can offer and deliver are extremely varied. For example, in a residential area the primary service the CO would deliver could be voice services. However, in a business district the CO may be more structured to support data and centrex-type services. However, with the advent of internet popularity many residential COs that primarily delivered voice services are now transitioning from a circuit-switched system to a packet-switching system.

A simplified example of a typical CO layout is shown in Fig. 1.6. Naturally, the dimensions and specific equipment required for the facility will need to factor in the type of services to be provided as well as the time frame the design is to encompass, i.e., growth.

The following is a brief list of the elements associated with a fixed network design that in turn is associated with a wireless system. There are obviously more elements to a network design, including coordination issues with adjacent markets as well as with the various vendors.

Typically, a CO consists of an equipment room, a toll room, a power room, and an operations room. The functions of each are unique in that the equipment room has the switching and packet platforms for treating and servicing the subscriber's needs. The toll room, also referred to as the interconnect and Telco room is the area of the MSC where the system interfaces to the PSTN, CLECs, IXCs, and other outside carriers. The purpose of the toll room is to provide the portal for entry and exit of services for the CO. The power room usually houses the rectifiers, batteries, and generators for emergency backup purposes. The operations center is the area where the craft personnel perform the data entry and monitoring and maintenance of the network itself.

The following list of topics should prove helpful in establishing the resources and timing needed for a fixed network design to be successful.

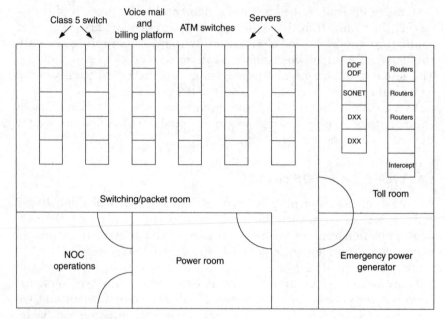

Figure 1.6   Typical CO layout.

Equipment room
- Class 5 switch
- Asynchronous Transfer Mode (ATM) switches
- Voice mail system
- Servers
- Billing system

Toll room
- Signaling Transfer Point (STP)
- Digital Cross Connect (DXX) equipment
- Routers
- Intercept equipment

The above does not address the issue of colocation with other service providers and the need to create a separate area for operators to maintain and upgrade their equipment.

1.5   Generic Cell Site Configuration

Figure 1.7 is an example of a generic cell site configuration. The cell site configuration shown in Fig. 1.7 is a picture of a monopole cell site. The monopole cell site has an equipment hut associated with it that houses

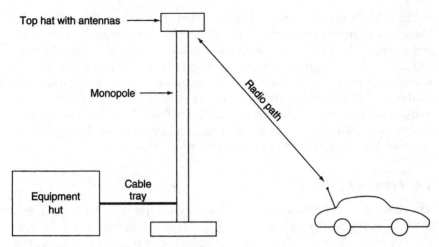

Figure 1.7 General cell site configuration.

the radio transmission equipment. The monopole, which is next to the equipment hut supports the antennas used for the cell site at the very top of the monopole. The cable tray which is between the equipment hut and the monopole supports the coaxial cables that connect the antennas to the radio transmission equipment.

The radio transmission equipment used for a cellular base station, located in the equipment room, is shown in Fig. 1.8. The equipment room layout is a typical arrangement in a cell site. The cell site radio equipment consists of a *Base Site Controller* (BSC), Radio Bay, and the Amplifier (TX) Bay. The cell site radio equipment is connected to the *Antenna Interface Frame*, AIF, which provides the receive and transmit filtering. The AIF is then connected to the antennas on the monopole through use of the coaxial cables that are located next to the AIF Bay.

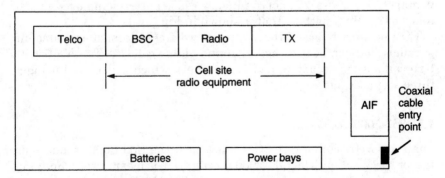

Figure 1.8 Cell site equipment room.

The cell site is also connected to the MTSO through the Telco Bay. The Telco Bay either provides the T1/E1 leased line or microwave radio link connection. The power for the cell site is secured through use of power bays, rectifiers, which convert ac electricity to dc. Batteries are used in the cell site in the event of a power disruption to ensure that the cell site continues to operate until power is restored or the batteries are exhausted. In many of the micro- and picocells the Telco interconnection is enclosed in the cabinet itself leading to a lower physical footprint for the site, thereby easing land-use acquisition issues.

1.6 Frequency Reuse

The concept and implementation of frequency reuse was an essential element in the quest for cellular systems to have a higher capacity per geographic area than an MTS or Improved MTS. Frequency reuse is the core concept defining a cellular system and involves reusing the same frequency in a system many times over. The ability to reuse the same radio frequency many times in a system is a result of managing the *carrier to interferer signal* levels, *C/I* for an analog system. Typically, the minimum *C/I* level designed for a cellular analog system is 17 dB *C/I*. However, for each different radio access technology used the *C/I* level will be different.

In order to improve the *C/I* ratio the reusing channel should be as far away from the serving site as possible so as to reduce the I component of *C/I*. The distance between reusing base stations is defined by a ratio referred to as the *D/R* ratio. The *D/R* ratio is a parameter used to define the reuse factor for a wireless system. The *D/R* ratio, Fig. 1.9, is the relationship between the reusing cell site and the radius of the serving cell sites. The table below illustrates standard *D/R* ratios for different frequency reuse patterns, *N*.

As the *D/R* ratio table (Table 1.1) implies, there are several frequency reuse patterns currently in use throughout the cellular industry. Each of the different frequency reuse patterns has its advantages and disadvantages. The most common frequency reuse pattern employed in cellular is the *N* = 7 pattern that is shown in Fig. 1.10.

The frequency repeat pattern ultimately defines the maximum amount of radios that can be assigned to an individual cell site. The *N* = 7 pattern can assign a maximum of 56 channels, which is deployed using a three-sector design.

1.7 IMT-2000

For a service to claim to be 3G it must meet the IMT-2000 standard or rather specification. The IMT-2000 specification is meant to be a unifying specification allowing for mobile and some fixed high-speed data services

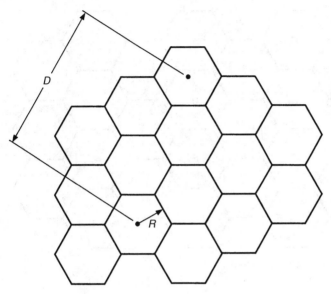

Figure 1.9 *D/R* ratio.

using one or several radio channels coupled with fixed network platforms for delivering the services envisioned.

- Global standard
- Compatibility of service within IMT-2000 and other fixed networks
- High quality
- Worldwide common frequency band
- Small terminals for worldwide use
- Worldwide roaming capability
- Multimedia application services and terminals

TABLE 1.1 *D/R* Rations versus Reuse Patterns

D	N
3.46	4
4.6 *R*	7
6 *R*	12
7.55 *R*	19

Note: *D/R* = distance to radius; *N* = reuse pattern.

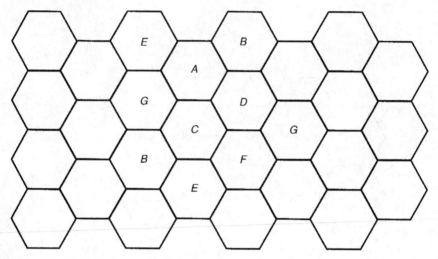

Figure 1.10 $N = 7$ frequency reuse pattern.

- Improved spectrum efficiency
- Flexibility for evolution to the next generation of wireless systems
- High-speed packet data rates
  - 2 Mbps for fixed environment
  - 384 Mbps for pedestrian
  - 144 kbps for vehicular traffic

Some of the more salient issues for 3G involve the interoperability and data rates. True interoperability between technologies is as elusive as the wireless data killer application.

There are multiple radio access platforms that comprise the IMT-2000 specification. Figure 1.11 shows the interrelationship between the various *radio access networks* (RANs) that are vying to be 3G and IMT2000 complaint.

1.8 Bluetooth

Bluetooth is a wireless protocol that operates in the 2.4 GHz ISM band allowing wireless connectivity between mobile phones, *Personal Digital Assistants* (PDAs), and other similar devices for the purpose of exchanging information between them. Bluetooth is meant to replace the infrared telemetry portion on mobile phones and PDAs enabling extended range and flexibility in addition to enhanced services.

Because Bluetooth systems utilize a radio link in the ISM band, there are several key advantages that this transport protocol can exploit. Bluetooth can effectively operate as an extension of a LAN or

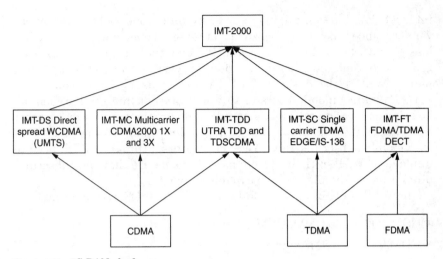

Figure 1.11   3G RAN platforms.

a Peer-to-Peer LAN (PAN) and provide connectivity between a mobile device and the following other device types.

- Printers
- PDAs
- Mobile phones
- LCD projectors
- Wireless LAN devices
- Notebooks and desktop PCs

One of the key attributes that Bluetooth offers is the range over which the system or connection can operate. Since Bluetooth operates in the 2.4 GHz *Industrial-Scientific-Medical* (ISM) band it has an effective range going from 10 to close to 100 m. The protocol does not require *Line of Site* (LOS) for establishing communication. Its pattern is omnidirectional, thereby eliminating orientation issues and can support both isochronous and asynchronous services paving the way for effective use of TCP/IP communication.

Bluetooth is meant to be a LAN extension fostering communication connection ease and not in delivering bandwidth.

1.9 WAP

Wireless Application Protocol (WAP) is one of the many broadband protocols being implemented in the wireless arena for the purpose of increasing mobility by providing mobile users the ability to surf the internet. WAP is being implemented by numerous mobile equipment vendors since it is meant to provide a universal open standard for wireless phones, i.e., cellular/GSM and PCS for the purpose of delivering internet content and other value-added services. Besides various mobile phones WAP is also designed for PDAs to also use this protocol.

WAP enables mobile users to surf the internet in a limited fashion, i.e., they can send and receive emails and surf the net in a text format only (normally you surf the net with graphics). For WAP to be utilized by a mobile subscriber the wireless operators, cellular or PCS, need to implement WAP in their system as well as ensure that the subscriber units, that is, the phones, are capable of WAP.

The WAP is utilized by the following cellular/PCS system types.

- GSM-900, GSM-1800, GSM-1900
- CDMA IS-95/CDMA2000
- TDMA IS-136
- UMTS (W-CDMA)

WAP increases the mobility of many subscribers and enables a host of data applications to be delivered for enhanced services to subscribers who use a smaller screen for displaying graphical information.

1.10   WLL

Wireless Local Loop (WLL) utilizes many similar, if not the same, platforms as used in cellular and PCS systems and is primarily focused on voice services. It, however, is different than cellular or PCS systems in its application which is fixed and not mobile. WLL being a fixed service is often referred to as *local multipoint distribution system* or *fixed wireless point-to-multipoint* LMDS or FWPMP. In fact WLL in many cases is the same as LMDS or FWPMP in its deployment and application. WLL is most applicable in areas where local phone service is not available or cost effective. Primarily WLL is a system that connects a subscriber to the local telephone company, PTSN or PTT, using a radio link as its transport media instead of copper wires.

There is no specific band that WLL systems occupy or will be deployed in. The systems can either operate in a dedicated and protected spectrum or use an unlicensed spectrum as their radio access method. Some of the services that fall within the definition of WLL include cordless phone systems, fixed cellular systems, and a variety of proprietary systems.

With the vast choices of system types and the spectrum considerations, the choice of which combination to use is directly dependant upon the application and services desired to be let. Some of the additional considerations for choosing the right technology platform involve the determination of the geographic area needed to be covered, the subscriber density, the usage volume and patterns expected from the subscribers, and the speed of deployment desired.

Since no single radio protocol and service can do everything, the choice of which system to deploy will be driven by the desired market and applications required to solve a particular set of issues. Some of the more common types of WLL systems involve Cellular, PCS, CT-2, and DECT, to mention a few of the common technology platforms.

WLL has different applications when deployed in a developed or emerging country. For a developed country the use of WLL takes on the premise of a cordless phone which is an extension of the house phone or PBX. For an emerging country the choice of WLL is more profound in that it is a cheaper alternative to that of laying wire by being more cost effective and is quicker to deploy. In many countries the use of cellular or PCS is quicker, easier, and cheaper to obtain than having a regular landline phone installed.

Figure 1.12 represents a typical WLL system. The WLL system has various nodes that are connected back to a main concentration point. The method for connecting the nodes to the concentration point can be either by radio, wire, or cable or a combination of all three. The introduction of 802.16 and 802.16a will provide an excellent bypass method for delivering WLL services.

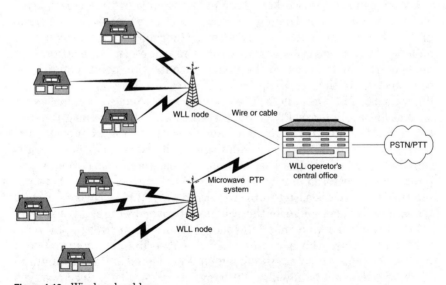

Figure 1.12 Wireless local loop.

1.11 LMDS

Local Multipoint Distribution System (LMDS) is a unique wireless access system that provides broadband access to multiple subscribers in the same geographic area. Point to multipoint is a concept where multiple subscribers can access the same radio platform utilizing both a multiplexing method as well as queuing. LMDS, while operating in the microwave frequency band and utilizing similar radio technology as a point-to-point microwave system, enables an operator to handle more subscribers, or rather Mbps/km^2, than a microwave point-to-point system using the same amount of radio frequency spectrum.

LMDS utilizes microwave radio as the fundamental transport medium and is not fundamentally a new technology but an adaptation of existing technology for a new service implementation. The new service implementation allows multiple users to access the same radio spectrum. LMDS is a wireless system that employs cellularlike design and reuse with the exception that there is no handoff. It can be argued that LMDS is in fact another variant to the WLL portfolio described previously and referenced as proprietary radio systems.

LMDS can be a very cost-effective alternative for a CLEC, *competitive local exchange carrier*. With LMDS a CLEC can deploy a wireless system without having to experience the heavy capital requirements of laying down cable or copper to reach the customers. The cost effectiveness is born out of the ability to focus the capital infrastructure where the customers are and at the same time being able to deploy the system in an extremely short period of time.

LMDS consists of two key elements—the physical transport layer and the service layer. The physical transport layer involves both radio and packet/circuit switching platforms. The radio platform consists of a series of base stations that provide the radio communication link between the customers and the main concentration point, usually the central office of the LMDS operator. Figure 1.13 is a high-level system diagram of an LMDS system. The system as shown has a similar layout as that for a cellular or PCS mobile system with the obvious difference being that the subscribers are fixed and operate at a different frequency.

The system diagram depicts multiple subscribers, customers, surrounding an LMDS hub or base station. The base station is normally configured as a sectored site for frequency reuse purposes and there are multiple subscribers assigned to any sector. The number of channels and the overall frequency plan for the system are driven by the spectrum available in any given market and the capacity required in any geographic zone.

LMDS enables the operator to over subscribe as compared to a microwave point-to-point system which is dedicated and not configured to share the same bandwidth with several subscribers. Specifically, a

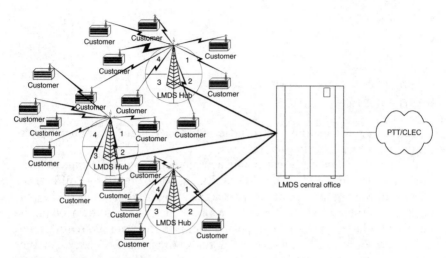

Figure 1.13 Generic LMDS system.

single radio channel may have 12 Mbps total throughput but one might be able to offer 24 Mbps or greater for the same channel by allocating it to the entire sector and not specific customers through overbooking. There are, of course, QOS issues and specific service delivery requirements with any commercial system. However, the concept is that an LMDS system utilizing point to multipoint technology can provide vastly greater bandwidth and services to a larger population than a point to point system can, utilizing the same spectrum.

Some of the services that LMDS can offer any given customer or customers are listed below for quick reference. It should be noted that the services listed are not all inclusive of what can be delivered or will be delivered. The only exception is that the service offered cannot have a bandwidth requirement greater than what the radio transport layer can support.The applications are as follows:

- LAN/WAN (VPN)
- T1/E1 replacement (clear and channelized)
- Fraction T1/E1 (clear and channelized)
- Frame relay
- Voice telephony (POTS and enhanced services)
- Video conferencing
- Internet connectivity
- Web services (email, hosting, virtual ISP, and the like)

- E-commerce
- VoIP
- FaxIP
- Long distance and international telephony
- ISDN (BRI and PRI)

A host of services and perturbations to those just listed above make an impressive portfolio to offer.

There are unique differences, however, in the service offerings allowed in Europe as compared to the United States. Some of the chief differences lie in different requirements or restrictions imposed on each of the perspective operators. Depending on the country the system is deployed in, the license will have restrictions imposing a minimum coverage requirement for universal services over a specified period of time, that is, 5 years. Another difference between LMDS in the United States and Europe is the spectrum assignments and the channel plans put forth by the European Community.

The EU operates in the 26 GHz band (24.5 to 26.5 GHz and 27.5 to 29.5 GHz) and has frequency blocks that are in divisions of 7 MHz and the channel allocation is usually done in increments of 28 MHz. The channel separation between the duplexed channel blocks for any 7 MHz channel grouping is 1008 MHz. There is also the 38 GHz band that is also being deployed, 37 to 39.5 GHz for systems in Europe. LMDS, however, is primarily associated with the 24, 28, and 38 GHz spectrum blocks each having a channel block of 50 MHz with four channels, each 12.5 MHz wide and duplexed.

The introduction of 802.16, discussed later in Chap. 7 promises to reduce or eliminate the proprietary systems leading to common RAN. The common RAN will enable mass production of the Customer Premise Equipment (CPE) that will make widespread deployment of LMDS economically feasible.

1.12   MMDS, MDS, and IFTS

Multichannel Multipoint Distribution Systems (MMDS), *Instructional Television Fixed Service* (IFTS), and *Multipoint Distribution Service* (MDS) are all sister bands to LMDS. The combination of MMDS, IFTS, and MDS bands make up what is referred to as *Wireless Cable*.

There are a total of 33 channels, each 6 MHz wide, that make up the MMDS, MDS, and IFTS bands collectively. The bands, while currently being referenced together, are all developed for different reasons. However, the bands were originally broadcast-related in that they were one-way oriented. The exception was the IFTS channel that has a part of the band allocated for upstream communication.

The band used by MMDS, MDS, and IFTS has numerous subscribers utilizing its service. However, there has been increased activity in redefining the services the band can and will offer subscribers. The primary focus of the band is toward high-speed internet traffic, IP, as compared to video services in conjunction with data. To make this happen the band(s) have been allocated for two-way communication, but the channels are not paired as it is done commonly in other bands. The two technology types which at this moment are competing for use in this band are Frequency Division Duplex (FDD) and Time Division Duplex (TDD) systems.

The technologies being deployed for MMDS, MDS, and IFTS bands is similar to LMDS in that it involves a sectored cell site that has multiple subscriber terminals associated with each channel, in every sector. One of the key advantages the MMDS, MDS, and IFTS bands have is the frequency these bands operate within. The bands for operation are in the range of 2.15 to 2.162 GHz and 2.5 to 2.686 GHz, which does not require strict adherence to line-of-site for communication reliability as well as the elimination of rain fade considerations in the link budget.

The chief disadvantage with this band is the coordination an operator must achieve in order to utilize a particular frequency in a geographic area. The coordination is exceptionally tricky due to the existence of MMDS, MDS, and IFTS operators that exist presently and primarily use video as their service offering. The coordination issue arises from both upstream and downstream frequency coordination since existing operators designed their systems based on a broadcast system.

However, this band is now defined for mobility and is an untapped resource at this writing; 802.16a, discussed in Chap. 7, addresses the MMDS band.

1.13 XDSL

XDSL is the term that is used to describe x-*type digital subscriber line* technology. There are numerous variants to Digital Subscriber Line technology types but they all have a similar premise and that is to convert the access line, twisted pair, into a high-speed data line allowing for a host of services to be offered with the existing infrastructure. DSL technology involves different modulation methods that enhance the data throughput capabilities of an existing access line, local loop.

Just which variant of DSL one may use is entirely dependant upon the application involved or the problem trying to be solved. Table 1.2 highlights the fundamental differences between the various DSL technologies.

The different types of DSL were listed in Table 1.3. Of the various forms of DSL used, *High Speed Digital Subscriber Line* (HDSL) and HDSL2 are used for delivering T1/E1 services using existing twisted

TABLE 1.2 DSL

DSL	Data rate	Comment
HDSL	1.544 Mbps or 2.048 Mbps	Symmetric, 2 pair
HDSL2	1.544 Mbps or 2.048 Mbps	Symmetric, 1pair
SDSL	768 kbps	Symmetrc, 1pair
ADSL	1.5–8 Mbps-down 16–640 kbps-up	Asymmetric,1pair
RDSL	1.5–8 Mbps-down 16–640 kbps-up	Asymmetric,1 pair, but changes data rate per line condition
CDSL	1 Mbps-down 16–128 kbps-up	Asymmetric, 1pair
IDSL	64 kbps	Symmetric, 1pair
VDSL	13–52 Mbps-down 1.5–6 Mbps-up	Asymmetric,1 pair

TABLE 1.3 Wireless Mobility Data

2G Technology	Data capability	Spectrum required	Comment
GSM	9.6 kbps or 14.4 kbps	200 kHz	Circuit switched data
IS-136	9.6 kbps	30 kHz	Circuit switched data
IDEN	9.6 kbps	25 kHz	Circuit switched data
CDMA (IS-95A/J-STD-008)	9.6 kbps/14.4 kbps 64 kbps (IS-95B)	1.25 MHz	Circuit switched data

2.5G Technology	Data capability	Spectrum required	Comment
HSCSD	28.8/56 kbps	200 kHz	Circuit/packet data
GPRS	128 kbps	200 kHz	Circuit/packet data
Edge	384 kbps	200 kHz	Circuit/packet data
CDMA2000-1XRTT	144 kbps	1.25 MHz	Circuit/packet data

3G Technology	Data capability	Spectrum required	Comment
WCDMA	144 kbps vehicular 384 kbps outdoors 2 Mbps indoors	5 MHz	Packet data
CDMA2000- EVDO/EVDV	144 kbps vehicular 384 kbps outdoors 2 Mbps indoors	1.25 MHz	Packet data

pairs. Synchronous Digital Subscriber Line (SDSL) may be used for delivering services in a multiple dwelling unit.

Asymmetric Digital Subscriber Line (ADSL) is interesting in that it is becoming a very popular residential service offering that is focusing on the lower end of the data market, the home user. ADSL has multiple business applications that can be used in conjunction with an LMDS operation. One such function involves having an ADSL service used by microcells used for both cellular/GSM/PCS sites in order to reduce monthly facility costs because with a partial T1/E1 being used, the bandwidth requirements in both directions fall within the bandwidth capabilities of a residential ADSL service thereby utilizing a residential service for a commercial application.

Additionally, ADSL could also be used to provide terrestrial connectivity for 802.11 AP backhaul, which will be discussed later.

1.14 Cable Systems

The proliferation of cable modems, primarily in the United States, has brought broadband to many end users who were previously relying on dial-up IP. Cable operators have a unique advantage for delivering broadband services, as do the Public Telephone and Telegraph (PTT), regarding the residential market in that they have a presence in many residential homes.

The common issue facing all broadband providers is the quality of their underlying transport layer. The quality of the cable plant itself dictates the delivery of services that can effectively be offered. The issues with the quality of the cable plant are primarily driven by the amount of drops that are on any cable leg which directly impacts the ingress noise problem that limits the ability for the cable plant to provide high-speed two way communication. Since most of the information flow is from the head-end to the subscriber, the system does not have to support symmetrical bandwidth requirements.

A *Hybrid Fiber/Coax* (HFC) network is shown in Fig. 1.14 with the enhancement of providing two way communication for both voice and data, besides the video service offering. The primary access method are physical media where the connection made to the subscriber at the end of the line is via coaxial cable. For increased distance and performance enhancements, fiber optic cables can be and are often part of the cable networks topology.

1.15 VoIP

Voice over IP has and continues to provide a viable alternative for call delivery for voice traffic. It is interesting that most of the initial VoIP

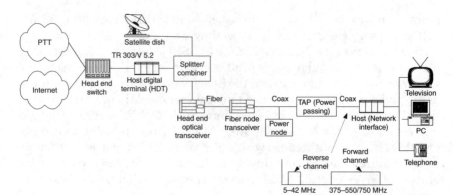

Figure 1.14   HFC.

implementations have not occurred over the Internet but rather over corporate LANs and private IP networks like *long distance* (LD) providers, which has mitigated the *quality of service* (QOS) problems associated with VoIP on the internet.

VoIP when mentioned in many circles invokes quality concerns due to delay and jitter problems, and thus QOS, when the access media are over the public internet. As mentioned previously, the true application for VoIP are transport media over private or dedicated pipes or networks where the QOS issue no longer is an issue.

The original standards activity for VoIP was defined in H.323 which has the name *Packet-Based Multimedia Communication Systems*. This standard's wide use was a direct result of offering it as freeware by Microsoft. There is, however, an alternative standard which is currently in competition with H.323 and that is *Media Gateway Control Protocol* (MGCP), also called *Single Gateway Control Protocol* (SGCP).

SGCP assumes a control architecture that has an architecture similar to the current PTT voice system where the control is done outside the gateway itself. The external call control elements are referred as call agents.

Wireless and CLEC operators that use infrastructure that is IP-based only can also offer voice services as part of their offering if the proper QOS and delivery issues are addressed in the design and service offering. For the wireless operators that offer voice via CES, voice services are an attractive entry point for customers. However, the discussion of VoIP does not need to be conveyed to the customer if the proper delivery and QOS issues are addressed.

A primary reason why VoIP is so attractive for a wireless operator is not solely related to the interconnect savings that may be achieved but in saving the spectrum since IP traffic is by itself dynamic bandwidth.

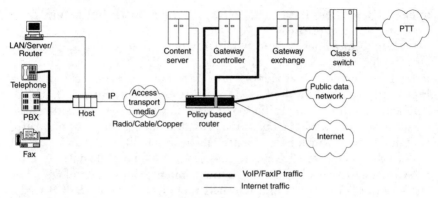

Figure 1.15 VoIP Network.

Figure 1.15 is a depiction of the major components involved with providing VoIP, either as a direct service or as an alternative transport medium that the wireless operator uses to be more cost competitive or better yet improve the margin.

In Fig. 1.15 VoIP can be delivered either directly to a public data network or the internet depending on the *Service Level Agreement* (SLA) that is used. In addition, the diagram depicts the issue of the operator using VoIP as a medium for handling voice traffic into the switching complex where it then converts the IP traffic into classical TDM traffic for interfacing to the PTT for call delivery.

1.16 Mobile Data (IP)

Mobile data are the future for wireless and is the thrust of this book. Mobile data have different meaning and requirements depending on what application or demand is being served.

Though mobile data have been used for decades, the need to extend the corporate LAN, surf the internet for information, use VoIP, or perform an IP to IP call (PTT) is just beginning to be felt. The hope for wireless data has been heralded for many years as just being around the corner. But the truth is that wireless data are unknown for mobility in terms of user acceptance and real network benefits from the operator's aspect, besides the customer.

A central concept need to be kept in the forefront when addressing mobile data, and that is that data applications are vertical. Vertical in that every user's specific data requirements are different. For instance, your individual data requirements may be casual internet browsing and IM capability. However, another user's requirement may be order management and FTP access. The point I am trying to address is that apparently wireless data seem to be rather simplistic but actually are profound and complicated.

Cellular Data Packet Data (CDPD) is one of the examples that the wireless industry needs to always remember. CDPD is and was an excellent packet network that rivals many of the current wireless data offerings and was available for close to a decade prior to 2.5G/3G. Applications that are specific to the user is what is needed, otherwise wireless operators will be only pipe providers facing constant price erosion due to bandwidth being treated as a commodity.

With that said there are several key elements for mobile data that are relevant for this book. There are many mobile data technologies but focus requires exclusion and wireless mobility for cellular/PCS is the focus. The key platforms that are used for wireless mobility are listed in Table 1.3.

The obvious missing component to the data offering is 802.11, whether it is 802.11b, g, or a.

1.17 Wireless LAN (802.11)

Wireless LAN (WLAN) is another wireless platform that enables various computers or separate LANS to be connected together into a LAN or a WAN. Big advantage is that WLAN-enabled devices do not need to be physically connected to any wired outlet enabling flexibility for location as shown in Fig. 1.16.

The convergence of 802.11 with wireless mobility has been described as the real killer application. The killer application for mobility is that it will truly allow the subscriber to take advantage of all the applications available on the WWW while at the office, home office, or on the road at some unknown location, provided of course there is coverage. The issue of security and provisioning to make this a reality is not a trivial matter if true transparency is desired with the intranet of a company by its sales and support staff.

There are several protocols that fall into the WLAN arena, not all of which are compatible with each other, leaving open the possibility of the formation of local islands. The most prevalent WLAN protocol is IEEE 802.11 but Bluetooth is also being referred to as a WLAN protocol. 802.11 is an IEEE specification which is made of several standards, some of the more prevalent are 802.11a, 802.11b (Wi-Fi), and 802.11g, to mention a few.

What is interesting is that 802.11a operates in the 5 GHz, UNII band, while 802.11b and 802.11g operate in the 2.4 GHz ISM (industrial, scientific, and medical) band along with Bluetooth that also operates in the 2.4 GHz ISM band. The 802.11g specifically is meant to increase the data rate to 54 Mbps while providing backward compatibility for 802.11b (Wi-Fi) equipment. What this means is that 802.11g equipment operating

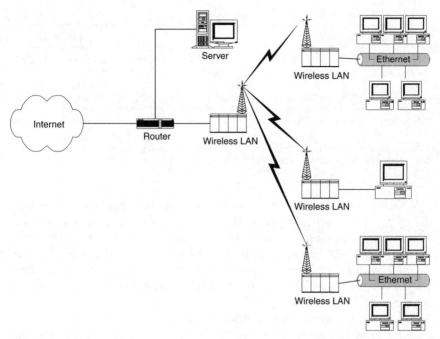

Figure 1.16   WLAN.

in the 2.4 GHz band can operate at speeds previously enjoyed by 802.11a equipment in the 5 GHz band.

To complicate matters there are a host of other 802 specifications, all of which either exist or are in the process of being standardized.

The 802.11 specification was designed initially as a wireless extension for a corporate LAN for enterprise applications and has numerous devices that have been manufactured to this specification. For example, the IEEE 802.11b protocol is a shared medium and utilizes CSMS/CA, which is a "listen-before-talk" protocol with standards for collision sense multiple access/collision avoidance.

The table that follows is a simple comparison between the key IEEE802.11 protocols and Bluetooth which is shown in Table 1.4. Both 802.11b and Bluetooth utilize the ISM band, but their format and purpose is different. However, 802.11a operates in the UNII band and can operate at a much greater ERP. Basically, the IEE802.11 devices are meant to cover a wider area then Bluetooth devices and the 802.11 devices have the potential of higher throughput. The data rate in the chart for IEEE802.11a and b shows a range in speeds which of course are dependent upon modulation format used, power, and also interference experienced.

TABLE 1.4 WLAN

WLAN	802.11a	802.11b	Bluetooth
Transport	5-Ghz UNII DSS	2.4-GHz ISM FHSS/DSS	2.4 GHz ISM FHSS
Data rate	6–54 Mbit/s	1–11 Mbit/s	1 Mbit/s
Range	*	50 m	1–10 m
Power	0.05/0.25/1 W	+20 dBm	0 dBm

*If used with an external antenna, the WLAN can be extended beyond the immediate office environment.

Note: DSS = direct sequence spread spectrum, FHSS = frequency-hopping spread spectrum, W = watts, dBm = decibels referenced to 1 milliwatt.

Why 802.11 is important for wireless mobility is because it provides direct mobile data interoperability between the local LAN of a corporation and the wireless operator's system. The inclusion of expending the corporate IP-PBX has great potential. Presently there have been many demonstrations and some operational systems regarding this integration of wireless mobility and wireless LANs which require the need for ASP's to enable the interoperability.

There is also another specification, HiperLan/2 which is a WLAN specification that has been developed under ETSI. HiperLan/2 has similar physical layer properties as 802.11a in that it uses OFDM and is deployed in the 5 GHz band. HiperLan MAC layers are different than 802.11a, hence, the different technology specification in that HiperLAN uses a TDMA format as compared to 802.11a which uses OFDM.

References

LaRocca, J., R. LaRocca, *801.11 Demystified,* McGraw-Hill, New York, 2002.
Smith, C., *LMDS,* McGraw-Hill, New York, 2000.
Smith, C., *Practical Cellular and PCS Design,* McGraw-Hill, New York, 1997.
Smith, C., D. Collins, *3G Wireless Networks,* New York, 2002.
Smith, C., C. Gervelis, *Cellular System Design and Optimization,* McGraw-Hill, New York,1996.
Winch, R., *Telecommunication Transmission Systems,* 2d ed., McGraw Hill, New York, 1998.

Radio System

The foundation of a wireless system whether it is mobile, fixed, or a hybrid is the *Radio Access Network*, RAN. There are numerous RANs that are used and the more relevant protocols that make up those RANs are covered in later chapters. However, regardless of the particular RAN protocol used, there are several common issues that transcend technology:

- Radio system
- Modulation
- Propagation
- Antennae
- Link budget

The preceding five topics list more physical attributes associated with the RAN. However, the fundamental underpinning of any successful design and network adjustment is based on the existence of engineering guidelines. Radio frequency (RF) engineering, both design- and performance-based, requires guidelines. The RF guidelines can be either formal or informal. However, with the level of complexity rising everyday in the wireless communication systems the lack of a clear definitive set of design guidelines is fraught with potential disaster. While this concept seems straightforward and simple, many wireless engineering departments have a difficult time defining their exact design guidelines when pushed.

This is especially true with regard to integrating various wireless platforms together due to the varying RF deployment and integration requirements. For instance, the deployment of a GSM/GPRS cell should have different design requirements than those used for deploying an 802.11 access point.

There are several very good sources for wireless mobility design guidelines related to the RAN. Some of the more pertinent sources for wireless mobility design and optimization are included in the references at the end of this chapter.

The aim of this book is to provide some fundamental information for the integration on various last-mile wireless access platforms into existing wireless mobility systems. This chapter will try to consolidate many of the more important issues with the generation and execution of the design criteria associated with the radio access portion of a system.

2.1 RF Design Process

The RF design process plays a critical role in the success of any wireless company, whether it is fixed or mobility-based. The RF design process used for integrating a 2.5G/3G wireless network with an 802.11, 802.16, or 802.20 system involves many common aspects that draw upon both microwave point-to-point and Cellular/PCS system designs. However, regardless of the technology platform used for the RAN the RF design guideline is a set of rules, criteria that are used to not only design the RAN and the new components that are added, i.e., cell sites, but also for improving the overall performance of the network. The design criteria should be structured so that the system will be configured to offer the best service within monetary and technological constraints.

Therefore, the design criteria for the radio access part of wireless mobility systems are extremely important to establish at the onset of the design, whether it is for a new system, for migrating to a new platform, or for expanding an existing system. There are many aspects associated with an RF design, and they are surprisingly common, in concept, with any radio access platform that is being used by wireless operators. The purpose of the design review is to ensure that there is an integrated design that takes on a holistic approach that spans across all the functional departments.

The overall RF design process is made of up of three fundamental areas or groupings. Within each design phase there are numerous subtasks or areas which can either be elaborated upon or eliminated depending on whether the design is related to a new system deployment, new service delivery within an existing system, or simply an expansion of an existing network.

A key concept is that the RF design process should be the fundamental structure used in the performance-improvement process. Using a design process, whether it is for hardware, software, or any potential alteration to a wireless system will only result in improvements, or, more importantly, the avoidance of major errors.

The RF design process can be broken down into three major components. Naturally, each component will have some differences depending on the objective for the design. The three design steps are as follows:

1. High-level design review (HDR)
2. Preliminary design review (PDR)
3. Critical design review (CDR)

The first step in the design process is the high-level design which is completed with the high-level design review (HDR). The actual high-level design and review involve multiple departments within the company. Specifically the high-level design review is meant to ensure that the company's marketing, financial, sales, customer care, and technical groups have an integrated approach to deploying a system or implementing a new service.

The RF engineering objective of the HDR is to create a formal review of the system solution for the radio environment aspect. Issues that will be addressed involve the interface, capacity, quality, and availability as compared to the company's marketing, financial, and sales objectives taking into account legal and regulatory requirements.

Some of the suggested RF design issues that need to be included in the HDR are as follows:

- Topology and architecture selected
- Available capacity and quality levels supported
- Technology choices
- Integration into existing network architecture
- Vendor and equipment/system selections
- Network management architecture (if applicable)
- Growth concepts
- System reliability concepts (i.e., redundancy and disaster recovery)
- Future services/platforms concept

The preliminary design and preliminary design review (PDR) process falls primarily under the technical service responsibility. This is a review of the various platforms involved with supporting the concepts and decisions made in the high-level design review. The purpose of the PDR is to identify and open issues or areas where critical decisions need to be made and discuss the alternatives. The PDR is meant to define the direction of the technical design.

The RAN design process—components for an existing wireless system using 2.5G/3G technology—is listed below.

1. Define plan (marketing/sales)

2. Identify access point target areas

3. Establish technology platform decisions

4. Establish *Serving GPRS Support Node* (SGSN)/ *Packet Data Serving Node* (PDSN) interaction (tightly/loosely coupled)

5. Determine subscriber usage

6. Determine coverage requirements

7. Establish environmental corrections (throughput, attenuation)

8. Determine maximum radius per AP (link budget)

9. Establish AP configuration and antenna system design

10. Establish the number of APs to cover area(s)

11. Generate coverage propagation plot for system and areas showing before and after coverage

12. Establish and define backhaul (802.16 or PSTN or existing 2.5G/3G backbone)

13. Check design against budget objective, and, if exceeded, re-evaluate design

It is important to realize that the list provided is not all inclusive and that it should be used as a foundation from which to alter.

The conclusion of the PDR allows the RF design team to continue with the design effort. If during the PDR the design or methodology required alteration due to design assumption flaws or changes to the business plan, the PDR may need to be redone or even the HDR.

Assuming that the PDR was successful, the next step in the design process is the *critical design* phase. The critical design is where the decisions made in the PDR are used to complete and refine the final design. The final design, referred to here as the critical design then undergoes a final design review called the *critical design review* (CDR). The CDR is that part of the design phase where no more changes are made to the design unless there is a flaw in the design process that requires the revisiting of critical design material resulting in a rescheduling of the CDR. The CDR should have the same format as the PDR in terms of the desired content with the exception that in the CDR design options are no longer relevant to the discussion since these decisions were made in the PDR.

Therefore, with the completion of the CDR the system design is completed. The design review process and subsequent revisions to the design itself will then be included in the quarterly or 6-month design review process that should be required by any wireless service company.

The overall design process flow is best depicted in Fig. 2.1.

2.2 Radio System

All wireless transport systems, whether they are fixed or mobile, have similar design issues. Obviously, there are technology-specific issues which will be identified in this chapter. However, to avoid duplication due to commonality issues, a general process will be described first prior to delving into more specifics. The commonalities for a design, however, cannot be overlooked since they, as it is commonly known to experienced engineers, play a major role in the design and functionality of any wireless system.

Since radio is a critical element in the successful deployment of a wireless system, a brief discussion regarding the physical components of a radio system needs to be included here. The RAN for any wireless system uses radio frequency spectrum as the transport mechanism. The fundamental building blocks of the radio aspect of the RAN should essentially be understood by the design and performance teams for the purpose of maximizing the overall customer experience with the final system.

The fundamental building blocks of a communication system are shown in Fig. 2.2.

The simplified drawing in Fig. 2.2 represents the major components in any wireless communication system from a radio aspect. The major components shown in Fig. 2.2 are an antenna, filters, receivers, transmitter, modulation, demodulation, and propagation. Each of the major components identified in Fig. 2.2 require the engineer to consider all the various perturbations in order to achieve the optimum design for the situation.

Entire books can and are written on each section listed above in Fig. 2.2. It is essential to know all the major components that actually make up the communication system that is being designed. By knowing the design characteristics of each of the components essential in building a communication system the proper transport functions for the information content can be achieved.

Figure 2.2 depicts the RF path from sender, transmitter, to receiver. However, this is just a part of the system that comprises the RAN itself. The RAN includes the physical radio elements themselves. However, depending on the technology deployed for achieving the access there are several variations that can and do take place for determining what exactly is included in the RAN. The definition of the RAN and what it

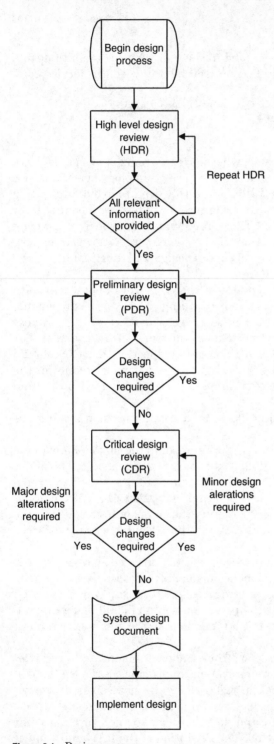

Figure 2.1 Design process.

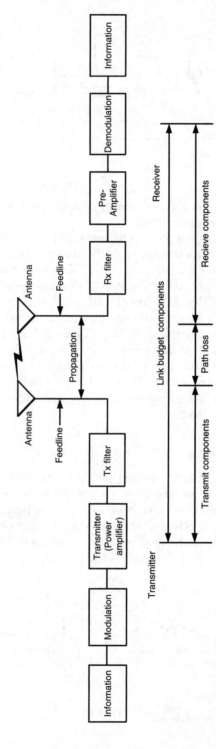

Figure 2.2 Communication system block diagram.

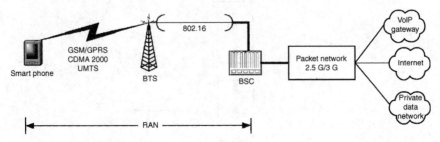

Figure 2.3   2.5G/3G RAN.

entails is important for determining bandwidth and coverage requirements, both of course are related.

Figures 2.3 and 2.4 show two slight variations of the components that
comprise a RAN. A RAN involves all the elements that are part of the
radio access network, by definition. The RAN elements are of course different for a 2.5G/3G and those for an 802.11 access point.

An examination of Fig. 2.3 shows that the RAN is associated with all
the elements that are used by the mobile subscriber to connect it to the
packet core network, SGSN, or PDSN. Figure 2.4, on the other hand,
indicates that the RAN is the connectivity used for the subscriber connecting to the 802.11 access point and not the backhaul used to connect
it to the packet network which can be both public and private. The RAN
for Fig. 2.4 includes the components which make up the 802.16a and
802.16 as part of the backhaul.

2.3   Propagation Model

The radio wavelength that is used for 2.5G/3G, 802.11, 802.16, and
802.20 is rather small in size and as a result has some unique propagation characteristics that have been modeled by numerous technical
people over the years. Some of the more popular propagation models
used are Hata, Carey, Longley-Rice, Bullington, and Cost 231.

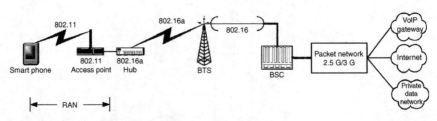

Figure 2.4   RAN - 802.11 access point.

TABLE 2.1 LOS and Non-LOS

Access technology	LOS/Non-LOS	Propagation model
2.5G/3G	Non-LOS	Hata and Cost231
802.11b,g (2.4GHz)	Hybrid	Cost231
802.16a	Hybrid	Cost231
802.16	LOS	Free Space
802.20	Non-LOS	Cost231

Each of these models has advantages and disadvantages associated with them. Specifically, there are some baseline assumptions used with any propagation model that need to be understood prior to using them. The most pressing issue is which propagation model you use.

The simple answer to the question is that "it depends." The specific propagation model used depends on whether the system is designed for *line of sight* (LOS) or non-LOS. 2.5G/3G wireless networks depend on non-LOS for their fundamental communication method while 802.16 requires LOS and 802.16a, depending the infrastructure used, could use non-Los as well as LOS. 802.11 is meant for non-Los but in many applications the fundamental principle of LOS applies. Table 2.1 shows a comparison for LOS and non-LOS by access technology.

The propagation model used for predicting coverage needs to factor into it a large number of variables that directly impact the actual RF coverage prediction of the site. The positive attributes affecting coverage are the receiver sensitivity, transmit power, antenna gain, and the antenna height above average terrain. The negative factors affecting coverage involve line loss, terrain loss, tree loss, building loss, electrical noise, natural noise, antenna pattern distortion, and antenna inefficiency to mention a few.

To date, no overall theoretical model has been established that explains all the variations encountered in the real world. However, as wireless mobility systems continue to grow, there is a growing reliance placed on the propagation prediction tools. The reliance on the propagation tool is intertwined in the daily operation of the system.

Presently, most of the propagation tools available use a variation of the Hata model or the COST231 model. More detailed discussions regarding the Hata and COST231 models can be found in the reference material. However, the foundation or rather reference point for all *path loss models* employed is free space.

The equation that is used for determining free space path loss is based on $1/R^2$ or 20 dB per decade path loss. It is shown in Eq. (2.1):

$$L_f = 32.4 + 20 \log_{10} R + 20 \log_{10} f_c \qquad (2.1)$$

where R = distance from cell site, km

f_c = transmit frequency, MHz

L_f = free space path loss, dB

The free space path loss equation has a constant value that is used for the air interface loss, a distance and frequency adjustment. Using some basic values the different path loss values can be determined for comparison with later models discussed.

Regarding 802.11 Access Nodes, the need to determine where to place the access point and where it will cover is of great concern for the design team. Usually time and available resources, besides cost of deployment, do not always allow for the establishment of good detailed testing. The fundamental approach that is used more often is to use the LOS method and when applicable add another AP where needed after the system goes online.

For propagation the free space has an $\alpha = 2.0$; however, 802.11 has an $\alpha = 3.3$ and this is used for determining path loss at 2.4 GHz. The LOS equation that can be used is:

$$PL = 32.4 + 33 \log_{10}(R) + EA \tag{2.2}$$

Where EA is *environmental attenuation* and R is in meters.

However, it is not always possible to know what the various obstructions or environmental attenuation corrections are for a given area. Therefore, a two-tiered propagation model is used that takes into account LOS for ranges less than 8 m and non-LOS for beyond 8 m within a building.

$$PL = 40.2 + 20 \log_{10}(R) \qquad R < 8 \text{ m} \tag{2.3}$$

$$PL = 58.5 + 33 \log 10(R/8) \qquad R > 8 \text{ m} \tag{2.4}$$

In practice, the LOS is the ruling factor for coverage which then is coupled with capacity throughput. As discussed later in subsequent chapters, the throughput of the system, whether it is mobility or 802.x, decreases as you move away from the source.

2.4 Path Clearance

In a radio communication path you either have LOS or Non-LOS. 2.5G/3G systems use non-LOS as a fundamental design. However, radio path clearance is an essential criterion for any point-to-point, point-to-multipoint, or mesh communication system. There are many different types of communication systems that can be used in support of a wireless mobility namely 802.11 and 802.16, besides terrestrial microwave point

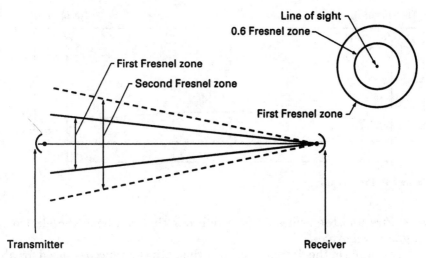

Figure 2.5   Fresnel zone.

to point. Depending on the specific technology platform used, path clearance may be required.

LOS or path clearance refers to the situation when the RF path cannot have any obstructions between the sending and receiving elements. Obviously LOS is associated with fixed wireless installations. An unobstructed LOS path is one that has no Fresnel zone violations.

The Fresnel zone is shown in Fig. 2.5. There are effectively an infinite number of Fresnel zones for any communication link. The Fresnel zone is a function of the frequency of operation for the communication link. The primary energy of the propagation wave is contained within the first Fresnel zone. The Fresnel zone is important for the path clearance analysis since it determines the effect of the wave bending on the path above the earth and the reflections caused by the earth's surface itself. Odd numbered Fresnel zones will reinforce the direct wave while even numbered Fresnel zones will cancel.

In a point-to-point communication system it is desirable to have at least a 0.6 first Fresnel zone clearance to achieve path attenuation approaching free space loss between the two antennae. The clearance criteria apply to all sides of the radio beam, not just the top and bottom portions represented by the drawing in Fig. 2.6.

Environmental effects on the propagation path have a direct influence on the point-to-point communication system. The environmental effects altering the communication system are foliage, atmospheric moisture, terrain, and antenna height of the transmitter and receiver.

The K factor used for point-to-point radio communication is 1.333 or 4/3 earth radius. The K factor ties in the relationship between the Earth's

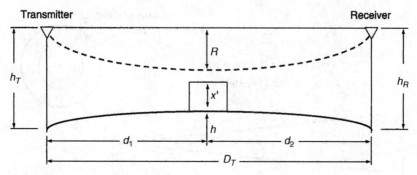

Figure 2.6 Path clearance.

curvature and the atmospheric conditions that can bend the electro-magnetic wave.

An example of the determination of the path clearance required for a point-to-point communication site is shown next.

Example To determine the path clearance required for a point-to-point communication site refer to Fig. 2.6.

First Freanel zone:

$$R = 72 \sqrt{\frac{d_1 d_2}{D_T f}}$$

Earth curvature:

$$h = \frac{d_1 \cdot d_2}{1.5\,k}$$

with f = frequency in GHz, d_1, d_2, and D_T in miles; x, h, h_T, and h_R in feet; and where $d_1 = 1.6$ mi, $d_2 = 2.1$ mi, $D_T = 3.7$ mi, and $f = 0.88$ GHz (or 880 MHz).

$$R = 72 \sqrt{\frac{(1.6)(2.1)}{(3.7)(.88)}} = \sqrt{\frac{3.36}{3.256}} = 73.14 \text{ ft} \quad R' = (0.6)R = 43.884$$

$$h = \frac{(1.6)(2.1)}{(1.5)(4/3)} = 1.68 \text{ ft} \quad k = \frac{4}{3}$$

Assume that the transmitter, receiver, and obstruction have the same ASML.

Earth curvature	1.68
0.6 Frenel zone	43.88
Obstruction height	100
	145.56 ft

The minimum T_x and R_x heights for the system are $h_T = h_R = 145.56$ ft.

2.5 Link Budget

The link budget used for any last mile access system is one of the key technical parameters used for the design process. Though the link budget directly determines the range and deployment pattern used, all the components in the design process are important. An important concept, especially when using multiple RANs, is that there can be several link budgets in any system based on the operating frequency, spectrum allocated, link reliability, physical components for the system, differences in up and down stream modulation methods or protocols, and rain fade issues, to mention some of the elements that need to be considered.

Additionally, the link budget decided upon for the system design needs to account for not only the physical issues defined but also marketing issues. Marketing issues address the path reliability deemed necessary for the market that is being targeted. For instance, if marketing determines that a link reliability of 99.99 percent is needed then adhering to 99.999 percent as a design makes little sense. Conversely, if a goal of 99.999 percent is desired for the service offering then designing the network for a 99.99 percent can result in a significant problem. The overdesigning of the network has advantages in terms of link reliability in that it builds in the added buffer for marketing changes in the future, based on outside competition requiring greater reliability. However, in the design process when presented by marketing a reliability number the radio link is only part of the reliability factor and this fundamental issue needs to be accounted for in the totality of the system design process.

The objective behind the link budget is to determine the path length or rather the size of the cell sites—access points—needed for the network design. The link budget is meant to accommodate multiple technology platforms and transverse over a wide range of radio spectrum issues. The link budget format presented here in Table 2.2 can and should be used when putting together the design of a system.

The above table captures most, if not all, of the important gain and loss components that comprise the link budget. The reason for conducting both uplink and downlink calculations is to establish the weakest link in the design path and for there set the cell radio requirements. In many access systems, the path does not require being balanced. It is only required that the path gain, with the appropriate link reliability value, be equal to or greater than the path loss determined.

The following relationships have to hold for the system to perform at the desired throughput level.

Downlink path gain >= path loss
Uplink path gain >= path loss

TABLE 2.2   Link Budget

Downlink		Units
Base station / AP Tx		dBm
Combining loss	( )	dB
Cable and connector loss	( )	dB
Antenna gain		dBi
Fade margin (99.999%)	( )	dB
Access point overlap	( )	dB
Client/AP Rx antenna gain		dBi
Noise figure (NF + cable loss)	( )	dB
C/N (Eb/No)	( )	dB
Rx sensitivity (x BER)		dBm
Downlink path gain		dB
Uplink		Units
Client/ AP terminal Tx		dBm
Cable and connector loss	( )	dB
Antenna gain		dBi
Fade margin (99.999%)	( )	dB
Base station/ AP Rx antenna gain		dBi
Noise gigure (NF + cable loss)	( )	dB
C/N (Eb/No)	( )	dB
Rx sensitivity (x BER)		dBm

Provided that both conditions above are met, the decision on which path is the limiting case needs to be made using the following method.

Downlink path gain – uplink path gain = X (dB)

Uplink path gain – downlink path gain = Y (dB)

You are to use the limiting case for the design that will either be the uplink or downlink path for an unbalanced system. In a balanced system the choice is irrelevant. However, in the preceding relationship, if Y is less than X, then the uplink path is the limiting design case.

The limiting path is then used with the path loss calculation to determine the coverage for the AP or site.

Therefore,

X (dB) > path loss (dB) + environmental corrections (dB)

Y (dB) > path loss (dB) + environmental corrections (dB)

As mentioned earlier, the throughput for the AP, as well as that for 2.5G/3G systems, will degrade as you move farther from the site. Therefore, it is important to determine the desired throughput and its relevant *Receive Signal Strength Intensity* (RSSI) value needed to ensure the required throughput desired.

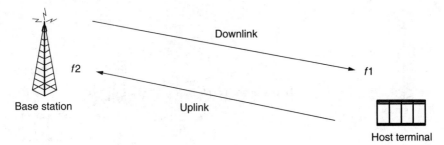

Figure 2.7 FDD system.

2.6 FDD and TDD

2.5G/3G and last mile access technology platforms use either a *Frequency Division Duplex* (FDD) or a *Time Division Duplex* (TDD) method for their RAN. Many of the deployed 2.5G/3G systems use an FDD system while 802.11 uses a TDD method. 802.16 uses both FDD and TDD depending on the application at hand (Fig. 2.7).

FDD uses two separate radio channels for communicating between the base station and the host terminal. One of the radio channels *f1* is for the communication link from the base station to the host terminal—downlink. The other radio channel *f2* is for the communication link from the host terminal to the base station—uplink.

The channels *f*1 and *f*2 are normally spaced some distance apart for isolation purposes. The FDD system uses a dedicated channel for uplink and downlink communication. Figure 2.8 illustrates how the uplink and downlink channels are paired.

A TDD system uses one radio channel for communication between the base station and the host terminal. The duplexing that is done is based on time and not frequency as is typically done with an FDD system. The

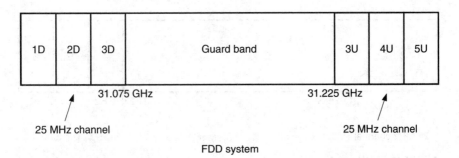

Figure 2.8 FDD spectrum allocation example.

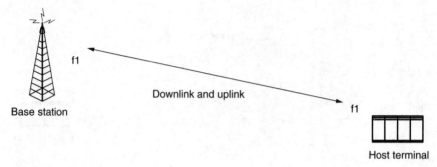

Figure 2.9   TDD system.

same channel $f1$ is used for both uplink and downlink communication between the base station and the host terminal as shown in Fig. 2.9. Figure 2.10 illustrates how the uplink and downlink channels are arranged.

TDD by its nature is designed to be more spectral efficient than an FDD system when involving data communication which is nonsymmetrical like Internet traffic. If the traffic is nonsymmetrical, as is IP Internet traffic where the downlink accounts for say 75 percent of the traffic, then four channels are used for TDD, which is the same as two channels for FDD.

The complexity with the TDD system versus an FDD system lies in the interference issues when the system is deployed in a mixed technology market, that is, TDD and FDD systems. The TDD system for both the base station and the host terminals needs to be coordinated with the FDD system since the host terminals will be transmitting when using a TDD on an FDD receive channel.

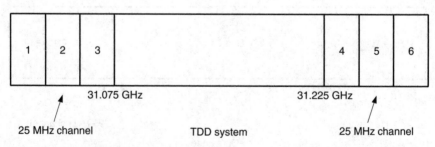

Figure 2.10   TDD spectrum allocation example.

2.7 Antennae

The antenna system for any radio communication platform used is one of the most critical and least understood parts of the system. The antenna system is the interface between the radio system and the external environment. The antenna system for a system can consist of a single antenna or multiple antennae at the base station and one at the host terminal station. Primarily the antenna is used by the base station site and host terminal for establishing and maintaining the communication link.

There are many types of antennae available, all of which perform specific functions depending on the application at hand. The type of antenna used in an AP system operator can be a dipole, path, or horn to mention a few. Coupled with the type of antenna is the notion of an active or passive antenna. The active antenna usually has some level of electronics associated with it to enhance its performance. The passive antenna is more of the classical type where no electronics is associated with its use and it simply consists entirely of passive elements.

Continuing with the type of antenna, there is the relative pattern of the antenna indicating in what direction the energy emitted or received from it will be directed. There are two primary classifications of antennae associated with directivity for a system and they are omni and directional. The omni antennae are used when the desire is to obtain a 360° radiation pattern. The directional antennae are used when a more refined pattern is desired. The directional pattern is what is used for 2.5G/3G systems with the variants being associated with the physical gain, aperture, its vertical and horizontal beam widths, and of course its polarization.

The choice of which antenna to use will directly impact the performance of either the base station, or the host terminal, or the overall network. The radio engineer is primarily concerned with the design phase of the base and host stations since these are fixed locations and there is some degree of control over the performance criteria that the engineer can exert on the base station and potentially on the host terminal location.

The correct antenna for the design can overcome coverage problems or other issues that are intended to be prevented or resolved. The antenna chosen for the application must take into account a multitude of design issues. Some of the issues that must be taken into account in the design phase involve the antennae gain, its antenna pattern, the interface or matching to the transmitter, the receiver used for the site, the bandwidth and frequency range over which the signals desired to be sent will be applicable, its power handling capabilities, and its IMD performance. Ultimately, the antenna you use for a network needs to match the system design objectives.

Table 2.3   dB

dB	Reference	Comment
dB	none	
dBm	1 mW	Standard wireless value
dBs	1 mW	Japanese wireless system reference
dBc	none	Referenced to the carrier power
dBw	1 W	1 W reference
dBk	1 KW	1 kw reference
dBu	1 mV	Standard wireless value
dBv	1 V	1 V reference

2.8 ERP and EIRP

Effective Radiated Power, ERP, and *Effective Isotropic Radiated Power*, EIRP, are the two most common references used for determining the transmit power of a communication site. ERP and EIRP are directly related to each other and a simple conversion can be achieved when one is known and the other is sought.

$$ERP = EIRP - 2.14$$

The following table can be used for a general comparison for various dB references. The dB reference Table 2.3 is not complete but covers most, if not all, of the dB references that are encountered in the wireless industry.

2.9 Modulation

To convey data and voice information from one location to another without physically connecting them together, as in wireless mobility, it is necessary to send the information in another way. There exist many methods for conveying information to and from locations that are not physically connected. Some of the methods involve talking and using flags, drums, and lights to mention a few. Each of the aforementioned methods has its advantages and disadvantages. However, the problem common with all the methods mentioned for communicating involves the physical distance the sender and the receiver have to be from each other along with the information throughput.

Electromagnetic waves are used to increase the distance and increase the information transfer rate between the sender and the receiver the use of an electromagnetic wave is made for this application. The use of electromagnetic waves is fundamental to radio communication. However, their use necessitates the modulation of the carrier wave at the transmitting source and then its demodulation at the receiver. The modulation and then demodulation of the carrier wave forms the principle of a radio communication system shown in Fig. 2.2.

The choice of modulation and demodulation used for the radio communication system is directly dependent on the information content desired to be sent, the available spectrum to convey the information, and the cost. The fundamental goal of modulating any signal is to obtain the maximum spectrum efficiency or rather information density per hertz.

There are many types of modulation and demodulation formats used for the transport of information. However, all the communication formats rely on one, two, or all three of the fundamental modulation types. The fundamental modulation techniques are *amplitude modulation* (AM), *frequency modulation* (FM), and *phase modulation* (PM). Figure 2.11 highlights the differences between the modulation techniques in terms of their impact on the electromagnetic wave itself.

$$E(t) = A \sin(2\pi f_c t + \phi) \tag{2.5}$$

where A = amplitude
    f_c = carrier frequency
    ϕ = phase
    t = time
    E = instantaneous electric field strength

Using Eq. 2.5, which can be used to define all types of modulation, the following are the key elements:

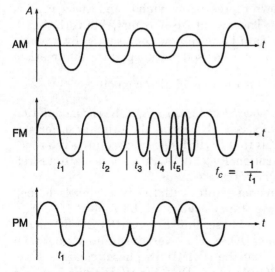

Figure 2.11 Modulation techniques.

- Amplitude modulation (AM) amplifies A
- Frequency modulaion (FM) modifies f_c
- Phase modulation (PM) modifies ϕ

Amplitude modulation has many unique qualities with it. However, this form of communication is not used directly in mobile wireless communication systems primarily because it is more susceptible to noise. A variant of AM, however, is used predominantly and that is *quardrature amplitude modulation* (QAM).

Frequency modulation is used for many mobile communication systems that employ analog communication. One common use of FM is in the analog Cellular communication since it is more robust to interference.

Phase modulation is used for conveying digital information. There are many variations to phase modulation, specifically many digital modulation techniques rely on modifying the RF carriers phase and amplitude as in the generation of QPSK and QAM signaling formats.

Quadrature Phase Shift Keying (QPSK) is one form of digital modulation that has a total of four unique phase states to represent data. The four phase states are arrived at through different I and Q values. Using four phase states, that is, quadrature, allows each phase state to represent two data bits. The two data bits are mapped on the I Q chart shown in Fig. 2.12.

The coordinate system for QPSK is best realized if you think in terms of an XY coordinate chart where X is now represented by the I, or in phase, and y is the quadrature portion, (I,Q). The distinct IQ location—phase state—shown represents a symbol, and this symbol is made up of two distinct bits. The advantage of using QPSK is the bandwidth efficiency. Since two data bits are now represented by a single symbol it requires less spectrum to be used to transport the information.

symbol rate = bit rate/(no. of bits/symbol)

Differential Quadrature Phase Shift Keying (DQPSK) is a modulation technique similar to that of QPSK. However, the primary difference between DQPSK and QPSK is that DQPSK does not require a reference from which to judge the transition. Instead, DQPSK's data pattern is referenced to the previous DQPSK's phase state.

DQPSK has four potential phase states with the data symbols defined relative to the previous phase state as shown in the Table 2.4.

Pi/4 Differential Quadrature Phase Shift Keying (Π/4 DQSK) modulation is very similar to that of DQSK. However, the difference between Π/4DQPSK and DQPSK is that the Π/4DQPSK phase transitions are rotated 45° from that of DQPSK. Like DQPSK, Π/4DQPSK has four

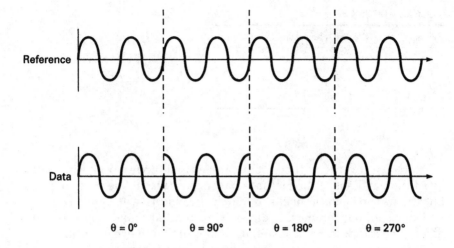

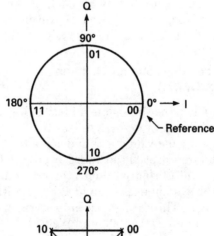

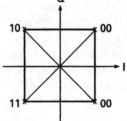

Figure 2.12   QPSK.

transition states and is defined relative to the previous phase state shown in the Table 2.5.

One method that is used to represent phase and amplitude modulation is through *I* and *Q* diagrams. The *IQ* diagram shown in Fig. 2.12 uses vector notation for representing the actual *I* and *Q* values.

TABLE 2.4   DQPSK

Symbol	DQPSK phase transition (degrees)
00	0
01	90
10	−90
11	180

There are several methods that can be used to view digitally modulated signals. Each of the methods has its positive and negative aspects but the method chosen needs to match the objective at hand. The four primary methods for viewing digitally modulated signals are spectrum display, vector diagram, constellation diagram, and eye diagram.

There are a multitude of text books and technical articles that abound the industry and focus purely on each of the modulation schemes if more in-depth analysis is required. However, one additional modulation format needs to be covered and that is OFDM.

2.10   Orthogonal Frequency Division Multiplexing (OFDM)

Orthogonal frequency division multiplexing is another modulation method that is used in many wireless fixed and mobile RANs. OFDM is a spread spectrum technique that distributes data over a large number of subcarriers that are spaced apart at precise frequencies, so they are independent and unrelated, hence orthogonal. Effectively, the radio bandwidth the system uses is divided into multiple carriers called subcarriers or tones. This form of spread spectrum is a different technique than that used for CDMA2000 and WCDMA.

OFDM is sometimes called multicarrier or discrete multitone modulation. OFDM is used in many applications ranging from digital TV to ADSL. But most important use of OFDM is for 802.11 and 802.16, to mention two applications related to mobile data.

TABLE 2.5   Π/4 DQPSK

Symbol	Π/4 DQPSK phase transition (degrees)
00	45
01	135
10	−45
11	−135

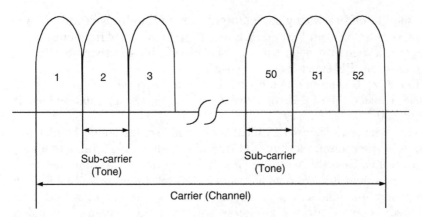

Figure 2.13 OFDM.

The 802.11a/g version of OFDM uses a combination of *binary phase shift keying* (BPSK), QPSK, and QAM, depending on the chosen data rate. It is important to note that 802.11b does not use OFDM.

There are a total of three 22 MHz channels with 802.11g and 8 to 25 MHz channels with 802.11a. The resulting OFDM carrier is 20-MHz wide and is broken up into 52 subcarriers, each approximately 300-KHz wide. OFDM uses four of these subcarriers as a reference to disregard frequency or phase shifts of the signal during transmission and they are used for error correction. The remaining 48 subcarriers provide separate wireless paths for transporting information in parallel.

Each subcarrier in the OFDM implementation is about 300-KHz wide (Fig. 2.13). The throughput is dependant on the modulation scheme used for each of the subcarriers in the OFDM signal. For instance, if BPSK is used to encode 125 Kbps of data per channel it results in a 6 Mbps data rate. However, if QPSK is used then the data per channel can be increased to 250 Kbps resulting in 12 Mbps data rate. The point to keep in mind, as with all wireless data RAN, is that the higher the data rate the more susceptible the signal will be to interference and fading, and ultimately the shorter the range, unless power output is increased.

2.11 Frequency Planning

Frequency planning or rather frequency management is an integral part of the wireless access system design. Frequency planning is a critical function for all wireless communication systems. How much frequency planning there is in a network is largely determined by the technology platform that is chosen by the operator for use. There are many variants to frequency planning ranging from coordination of a single transmit channel to orchestrating the manipulation of hundreds of radio chan-nels. Within the Cellular and PCS arenas the amount of frequency plan-ning can range from segmentation of the available spectrum to defining the different PN short codes for CDMA.

For 802.16 systems, the choice of frequency assignment methods varies based on the spectrum available as well as the modulation format employed.

There are a number of technical books and articles, some of which are listed in the references that deal with the frequency planning of a net-work from a theoretical standpoint. It is very important to understand the fundamental principles of frequency planning in order to design a frequency management plan for any network. Failure to adhere to a defined frequency design guideline will limit the system's expansion capability.

There are several methods available for use in defining the fre-quency management of a network. The method chosen by the mobile carrier needs to be factored into the frequency management scheme capacity requirements, capital outlays, and adjacent market integra-tion issues to mention a few. Obviously, the method that is used for the frequency plan also has to ensure that the best possible C/I ratio is obtained for both co-channel and adjacent channel RF interference.

The use of a grid is essential for initial planning, but, when "sprin-kled with reality," the notion is academic in nature. The reason it is aca-demic is because the site acquisition and physical implementation process tends to drive the system configuration and not the other way around. It is dealing with the irregularities of the site coverage, traffic loading, and configurations that require continued maintenance of the network frequency plan.

The specific frequency assignment scheme is directly related to the D/R ratio. The D/R ratio is the relationship between the radius of the serv-ice site and the distance to the next site that will reuse the same radio frequencies. Figure 2.14 shows the D/R relationship. There are multiple methods used to improve (reduce) the D/R ratio thereby maximizing the spectrum utilization.

As mentioned before, there are several different methods of assign-ing frequencies in a network. The most common methods used in wireless mobility are $N = 7$ and $N = 4$ (Fig. 2.15). The $N = 7$ method of channel assignments is one of the most popular methods for assign-ing frequencies. This pattern is shown in Fig. 2.16. There are several advantages with using an $N = 7$ pattern. The advantages lie in its ability to provide a high level of trunking efficiency, increase flexi-bility for the placement of the RF channels in the network, and excel-lent traffic capacity. The $N = 7$ pattern uses a 120° sectored cell design that equates to three sectors per cell site. This design facilitates the

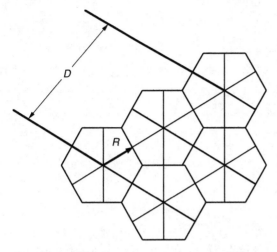

Figure 2.14 *D/R* ratio.

reuse of the same channels in close proximity to one another cell site.

The $N = 4$ frequency assignment method is another popular method for assigning frequencies. This pattern is shown in Fig. 2.16. The key advantage with using an $N = 4$ pattern is the high traffic capacity handling ability obtained. The $N = 4$ pattern delivers the maximum number of channels per square mile or kilometer out of all the channel assignment methods. However, the disadvantage with this method is that it is not as flexible for cell site placement and does not have the trunking efficiency under average traffic loads as the $N = 7$ pattern. The $N = 4$ pattern uses 60° sectors, six sectors per cell site for AMPS and IS-136 systems, but it is also used with 120° sectors with GSM.

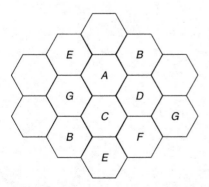

Figure 2.15 $N = 7$ frequency reuse.

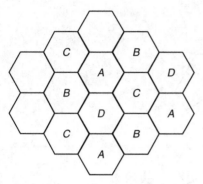

Figure 2.16 $N = 4$ frequency reuse pattern.

The $N = 3$ pattern shown in Fig. 2.17 is a frequency reuse scheme that should be implemented with the deployment of 802.11 into a wireless mobility system, or even as a separate multiple access point system. The frequency assignment scheme is based on the three nonoverlapping channels available with 802.11b/g systems.

It should be stressed that the frequency patterns used for an 802.11 deployment are independent of the overall wireless mobility system due to the fundamental concept that they occupy different parts of the frequency band and therefore, with the exception of the packet core network and the CPE, are separate systems that could have no interaction between them.

Figure 2.18 is a frequency assignment scheme that can be used for an 802.16a system that uses the MMDS frequency band. It is important to note that the use of horizontal and not just vertical polarization is made

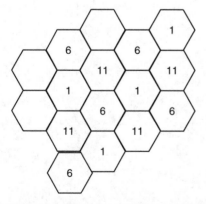

Figure 2.17 $N = 3$ frequency reuse pattern for 802.11.

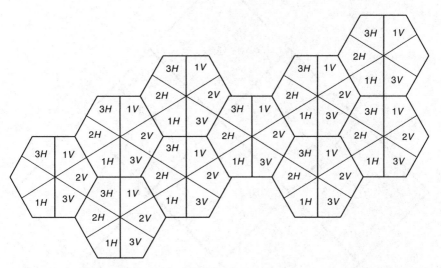

Figure 2.18   Deployment of three-channel 60° sectors (802.16a), using two polarizations.

for maximizing the spectrum utilization. The channel assignment method could be for TDD as well as FDD system configurations.

Figure 2.19, however, is another example of possible fixed wireless deployment schemes that use 802.16 LMDS band channels. The deployment scheme showed implies that only two channels are available for use and that they have the same polarization. The example shown in Fig. 2.20 is but one possible situation and the reference section of this book provides a source for an LMDS design and deployment guide.

Regardless of the technology platform used, the main idea to keep in mind when frequency managing a system is to maximize the distance between reusers; it's that simple a concept. Whatever method or process is used when maximizing the reuse distance, it will be fraught with problems. When selecting a frequency set for a cell site—whether it is a new site, expansion issue, or correcting an interference problem—the current and future configuration of the network needs to be factored into the equation.

The radio assignment planning methodology has several steps, which, again, are almost radio access platform independent.

Rules for assigning radio channels

1. Do not assign co-channels or adjacent channels at the same cell site (NA for CDMA).

2. Do not assign co-channels in adjacent cell sites (NA for CDMA).

3. Do not mix and match frequency assignment groups in a cell or sector.

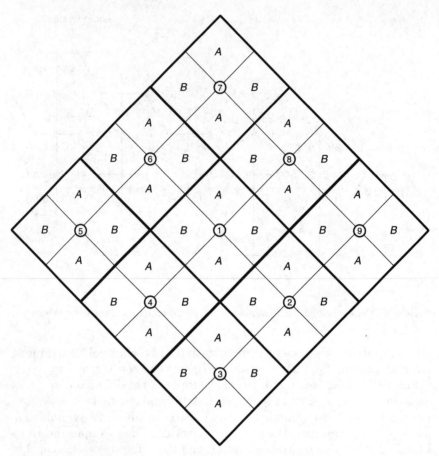

Figure 2.19 Same polarization, two-channel 802.16 deployment.

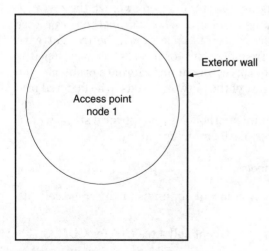

Figure 2.20 Single access point.

4. Avoid adjacent channels' assignments in adjacent cell sites (NA for CDMA).

5. Maintain proper channel spacing for any channel assignments for a sector or site.

6. Maximize the distance between reusing cell sites.

Obviously, the assignment rules can and should be elaborated upon, but the concept is to establish a set of rules, usually available from the infrastructure vendor, and follow them.

When putting together a frequency plan for a site or a system it is essential that a process exists that will ensure that the major items are always reviewed. The use of a check list will expedite the review process and ensure the accuracy of the plan.

2.12   Inbuilding and Tunnel Systems

Designing an interior wireless system to support an inbuilding, subway, or tunnel application can and should be a rewarding situation. Some of the applications include improving coverage for a convention center or large client, disaster recovery, and a wireless PBX. The issues that are encountered for interior communication system applications are unique to those associated with a macrosystem.

There has been a lot of focus on inbuilding applications over the years with surges in effort taking place. But the fundamental problem that keeps arising is the inherent cost of the facility. The cost of the inbuilding system far outweighs any potential for increased wireless revenue for the wireless operator. Cable and antenna installations, whether distributed antennae or leaky feeder, are the real driving costs of these installations. The inclusion of 802.11g wireless LANs has proved exceptional for configuration and installation ease due the proliferation of the 802.11 capability in most devices.

Typically, the inbuilding design approach has been the surround-and-drown method, which should be evident in the link budget calculations associated with urban environments. The propagation of the radio frequency energy, however, takes on unique characteristics in an interior or confined application as compared to an outdoor environment. The surround-and-drown method usually is the first step in helping define where possible customer usage could be followed by more surgical installations.

The advent of Wi-Fi hotspots, however, has greatly altered the equation for determining where the customer requirements will be since an access point node is relatively inexpensive to buy and install. The true cost of ownership for the access point is the backhaul required for connectivity to the Internet or a private data network.

For a Wi-Fi hotspot or mobile data, requirements in a building have unique design considerations which manifest themselves in the propagation of the signal itself. The primary difference in propagation characteristics for interior versus outdoor is the fading, shadowing, and interference. The fading situation for interior situations results in a deeper fade which is spatially closer than normally encountered in an exterior application. Shadowing is also quite different due to the lower antenna heights, excessive losses through floors, walls, and cubicles, and vehicle blockage as in the case of a tunnel. The shadowing effects severely limit the effective coverage area to almost LOS for wireless communication whether it is in the Cellular, PCS, or 802.11 band.

The interference issue with an interior communication system can actually be a benefit since the interference is primarily noise driven and is not co-channel interference, assuming no reuse for involving the interior applications. Obviously this is not a CDMA issue. For 802.11 systems portable phones should be selected so that they operate in the 900-Mhz band and not the 2.4-GHz band.

There are some unique considerations that must be taken into account regarding interior communication system designs. Some of the design considerations that need to be factored into the design are as follows:

1. Base or access point to subscriber unit power

2. Subscriber (CPE) to base/AP power

3. Link budget

4. Coverage area

5. Antenna system type and placement

6. Frequency planning

7. Throughput

The base to subscriber power needs to be carefully considered to ensure that the desired coverage is met, deep fades are mitigated in the area of concern, the amplifier is not being over or potentially under drive, and mobile overload does not take place. The desired coverage that the interior system is to provide might require several transmitters because of the limited output power available from the units themselves. The power limitation often makes the forward link the limiting path in the communication system for an inbuilding system.

The subscriber to base power also needs to be factored into the interior design. Imbalances could exist in the talk out to talk back path resulting in poor link quality. Poor link quality then equates to reduced throughput and overall degradation of the customer experience.

The link budget for the communication system needs to be calculated in advance to ensure that both the forward and reverse links are set properly. The link budget analysis plays a very important role in determining where to place the antenna system, distributed or leaky feeder, and the number of microcell/picocell systems required to meet the coverage. The link budget associated with an inbuilding system is for all intents and purposes a Line-of-Sight model. The simple rule is if you can see the antenna you have coverage. Interior fading and attenuation is very severe and rounding a corner in an office will usually result in a loss of signal which will deteriorate the call quality or have it terminate prematurely.

Reiterating the concepts for inbuilding or tunnel coverage rely on LOS and not multipath to ensure a reliable communication link. This is a fundamental change from the macrosystem design which by default relied on the use of multipath to ensure the communication link.

Figure 2.20 is an example of a possible Wi-Fi hotspot that has a single access point associated with it. The particular backhaul, authentication method, and services are not defined. The single AP could be on the 12th floor of a building or in the main lobby area used by visiting business people.

Figure 2.21 shows a similar situation as that depicted in Fig. 2.20 with the exception that multiple access points are deployed for the same area. The multiple access points could be deployed for improved coverage or throughput or both.

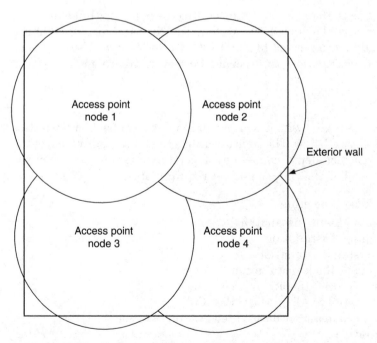

Figure 2.21   Multiple AP.

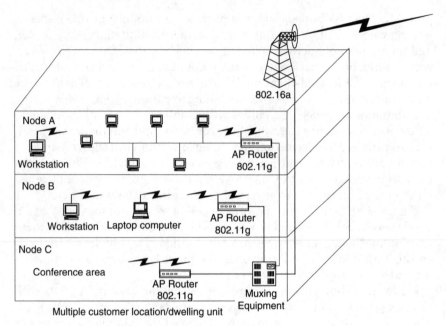

Figure 2.22 Multiple APs.

Figure 2.22 is an example of multiple access points within a building that are connected to an 802.16a system used for backhaul. The 802.16a system could be connected to a 2.5G/3G wireless network's cell site or any other concentration point needed for service treatment.

2.13 Planning

The issues associated with designing an inbuilding or backhaul wireless system are listed below. The time durations that accompany each of the steps are not included because they depend directly on the size of the system as well as the time to market requirements.

Project kickoff meeting
- Antenna Mount Installation
- Equipment Installation
- UPS System Installation
- Equipment Rack Installation
- AP or Microcell Install
- Cable Install to AP or Micro/Pico Cell
- Radiax Cable and or distributed antenna system
- Installation

RF engineering project design kickoff meeting
- Establishment of responsibility centers
- Performance criteria
- Review system designs
  - System review
  - Coverage requirements
  - Handover
  - Cable design
  - Performance criteria
  - Parameter adjustments
  - Facilities
  - ATP signoff

2.14 Intelligent Antennae

Intelligent antenna systems are being introduced to commercial wireless communication systems. The concepts and implementation of intelligent antenna systems have been used in other industries for some time, primarily in the military. The brief discussion regarding intelligent antennae is separate from the antenna section previously covered in this chapter, primarily to avoid commingling concepts.

Intelligent antenna systems can be configured as either receive only or full duplex operations. The configuration of the intelligent antenna systems affords them to be arranged as either in an omni or sector cell site depending on the application at hand.

The fundamental objective is to increase the S/N by reducing the amount of N, noise and interference, and possibly increasing the S, serving signal, in the same process. All the technologies referred to are based on the principle that narrower radiation beam pattern will provide increased gain and can be directed toward the subscriber and at the same time offer less gain to interfering signals that will arrive at an off axis angle due to the reduced beam width.

A simple rule of thumb that can be used for determining the amount of improved S/N that can be achieved with an intelligent antenna system, is shown in Fig. 2.23.

Improvement expected = $10 \log_{10}(4)$ = 6 dB, that is, assuming the sector is divided into four even pieces from its original configuration. More specifically, if the sector was 120°, a 6-dB improvement would be expected were the sectors reduced to 30°.

Figure 2.24 illustrates three types of intelligent antenna systems, each has positive and negative attributes.

All of the illustrations shown can be either receive only or full duplex. The difference between the receive only and the full duplex systems

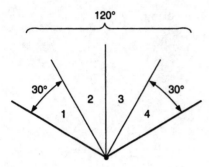

Figure 2.23 Improvement.

involves the number of antennae and the potential number of transmitting elements in the cell site itself.

The beam-switching antenna arrangement shown is the simplest to implement. It normally involves four standard antennae of narrow azimuth beam width, 30° for a 120° sector, and based on the receive signal received the appropriate antenna will be selected by the base station controller for use in the receive path.

The multiple beam array shown involves using an antenna matrix to accomplish the beam switching.

The beam steering array, however, uses phase shifting to direct the beam toward the subscriber unit. The direction that is chosen by the system for directing the beam will affect the entire sector. Normally amplifiers, but for transmit and receive, are located in conjunction with the antenna itself. In addition, the phase shifters are located directly behind each antenna element. The objective of placing the electronics

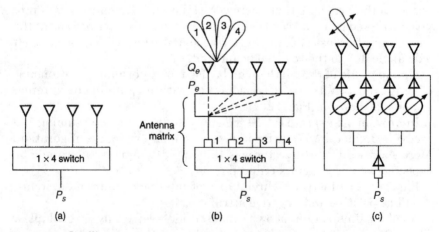

Figure 2.24 Intelligent antenna systems.

in the masthead is to maximize the receive sensitivity and exploit the maximum transmit power for the site.

References

Bing, B., *Wireless Local Area Networks,* Wiley, New York, 2002.

Carlson, A. B., *Communications Systems,* 2d ed., McGraw Hill, New York, 1975.

Jakes, W. C., *Microwave Mobile Communications,* IEEE Press, New York, 1974.

Johnson, R. C., and H. Jasik, *Antenna Engineering Handbook,* 2d ed., McGraw-Hill, New York, 1984.

Kaufman, M., and A.H. Seidman, *Handbook of Electronics Calculations,* 2d ed., McGraw-Hill, New York, 1988.

Lathi, B.P., *Modern Digital and Analog Communication Systems,* CBS College Printing, New York, 1983.

Lee, W.C.Y, *Mobile Cellular Telecommunications Systems,* 2d ed., McGraw-Hill, New York, 1996.

Minoli, D., *Hotspot Networks,* McGraw Hill, New York, 2003.

Rappaport, T., *Wireless Communications Principles and Practices,* IEEE, New York, 1996.

Reid, N., and R. Seide, *802.11 (Wi-Fi) Networking Handbook,* McGraw Hill, New York, 2003.

Smith, C., *LMDS,* McGraw-Hill, New York, 2000.

Smith, C., *Practical Cellular and PCS Design,* McGraw-Hill, New York, 1997.

Smith, C., and C. Gervelis, *Wireless Network Performance Handbook,* McGraw Hill, New York, 2003.

Reference Data for Radio Engineers, 6th ed., Sams, New York, 1983.

Chapter

3

Network Design

The wireless data network system design is unique for wireless systems in that it may, depending on the services offered and the technology platform chosen, involve multiple protocols to deal with. The use of multiple protocols within telephony and wireless systems is not unique. However, the fundamental choice as to which platform will be the predominant force in the network design can and does lead to many perplexing situations.

The complication is further magnified based on the amount of on-net and off-net traffic the system will need to handle and the types of protocols they comprise. The backhaul from the access point and/or base station to switching/packet concentration node further adds another wrinkle in this effort based on traffic volume and interconnect facilities that are available either in the time frame desired or the cost of using those facilities or both.

The telecommunication industry is moving toward convergence of the plethora of service protocols. The choice of the protocol to be used, migrated, has yet to be determined. Legacy services will also need to be supported until convergence really can take place. Convergence is occurring on the fixed access network as well as the RAN. The convergence in the fixed access network involves *Time Division Multiplex* (TDM) and packet data where packet data are *Asynchronous Transfer Mode* (ATM) and IP. Both ATM and IP have their advantages and are used for different applications.

So the perplexing question is, "Which platform do you use that will be 'future proof' and not require additional capital investments in the future due to technology obsolescence?" The answer to that question is it depends on which part of the network you are referring to. If the part of the network is the edge then the convergence is toward IP. However, for the core of the network the convergence is toward ATM. As always is the case, the solution is not based on a single killer protocol but the proper application of each toward obtaining the desired solution.

A decision will need to be made regarding the packet and circuit-switching network. In particular the decision will need to be made whether to lease capacity from another provider, thereby expediting the time to market and reducing operating expenses, but at the cost of control.

However, the most important decision that drives all the others relates to the services you intend offering either at system launch or in one or two years from now. The decisions made will dictate the network configuration, which, if not chosen well, will result in excess capital expenditures in the future to compensate for an incorrect decision.

Since there are multiple platform decisions to make, this chapter will attempt to cover the issues the network designer will need to address or at least consider in the design process.

3.1 Service Treatments

Wireless data services offered need to be realized. One of the areas where realization takes place is with the network design. The network engineering design is interested in the type of services offered, where they originate from, where they terminate, if any treatment is needed, the provisioning and monitoring methods, and of course the *committed information rate* (CIR) with an overbooking factor that will be promoted with each service.

What services can be offered depends largely upon the radio infrastructure and not the network engineering design. What I mean is that the network engineering design, being fixed, can support and deliver any protocol and service offering requested given the time and resources to accomplish the task. However, if the RAN interface to the customer is able to provide only 15 kbps of UBR traffic then offering rt-VBR for video conferencing is not viable because of the RF environment and not the fixed network.

Supporting the services and what that means is again a vast area that has multiple meanings. For instance, the service offering could simply be wireless operator offering connectivity and no content, that is, a pipe provider. Another example could be a service provider offering a LAN and *Internet Protocol Private Branch Exchange* (IP-PBX) extension. But what exactly is service? Is service the delivery of the bandwidth only or do you provide adjunct services to support the primary bandwidth provision, that is, sell *Customer Premise Equipment* (CPE), cable the customer, configure their routers, provide *Application Specific Programs* (ASPs), and so on?

Often the services offered will change with time due to the varying market conditions brought on by competition. Therefore, the platforms used by the network engineering design need to be flexible enough in design to account for the vast array of unknown changes. A seemingly daunting task but in reality the issue is really a scaling issue.

The three major platforms that need to be supported are

- ATM
- IP
- TDM

With each of these platforms the types of services that are to be offered will be proposed by the marketing department in conjunction with assistance from the technical community. Associated with each service offered, the following five primary topics need to be answered for each, more specifically what is the

1. Committed information rate (CIR)
2. Peak information rate (PIR)
3. Quality of service (QOS)
4. Overbooking factor
5. Fixed or variable billing

The above five questions are really derivatives of the services offered, with the noted exception of the overbooking factor.

However, the direction that the industry is migrating to is IP at the end terminal, i.e., mobile phone, with ATM as the backbone. The use of TDM circuits has to be considered in the interconnection method when connecting to another service provider for further delivery options. However, IP is the prevalent platform that will govern mobile data deployments.

3.2 TDM/IP/ATM Considerations

Well, how do you decide which platform to use, TDM, IP, or ATM and associated with this the dimension or proportion that each will be within the network? The answer of course is not simple since experience indicates that no one platform is a solution for all situations and requirements.

There are several things to consider when selecting the platform used for the fixed network layout.

1. What services do you need to support, keeping in mind that the key to mobile data success is targeting a niche market and not providing a general service to the customer with no sticky applications?

TABLE 3.1 Protocol Efficiency Comparison

Protocol	TDM circuit	Voice	Packet data	Video
ATM	1st	1st	3rd	1st
IP	NA	3rd	2nd	2nd
Frame Relay	NA	2nd	1st	NA

2. What is the RAN protocol and its bandwidth delivery capability?

3. What is the bandwidth required for connecting the base station or access point back to the concentration node, i.e., MSC?

4. Can wireless be used instead of a *Public Switched Telephone Network* (PSTN)/*Competitive Location Exchange* (CLEC) for connectivity between the *Base Transceiver Station* (BTS) and the MSC?

5. What is the reliability required between the various RAN and the MSC?

Table 3.1 indicates the types of platforms, i.e., the protocols that are best used for different types of services offered.

In Table 3.1 NA refers to the fact that the protocol does not have a standard method of supporting that particular service.

A fundamental difference ATM has versus IP and Frame Relay is that ATM uses fixed length cells while IP and Frame Relay use variable length packets.

Of course, TDM can handle ATM, IP, and Frame Relay services but not efficiently, i.e., cost effectively, in addition to realizing that unless you want to become the next PTT, which is your biggest competitor, building a TDM system may not be a desired solution.

With that said, voice services are still a large percentage of the wireless telecommunication market. Therefore, the use of class 5 switches is still prevalent throughout the industry. However, the reliance on class 5 switches is being supplemented by the use of soft switches.

3.3 TDM Switching

The role of the TDM switch in the network has grown along with the importance and complexity of the network itself. The exact role TDM switches will have in the future is unclear but they play an important role in wireless mobility systems today. The industry migration from TDM to IP/ATM platforms should seriously be considered when moving forward with the TDM switching decision.

However, no matter what is said, TDM switching will continue to play a major role in wireless mobility. TDM switches began their role with first the manual exchange (also referred to as just a switch), followed by the rotary exchange, the crossbar exchange, and eventually lead up to the development of the modern electronic *stored program control* (SPC) exchange. Even amongst the newer switches there are various designs and functional applications depending on the switch manufacturer and the specific role the switch serves in the network, i.e., central office, tandem, *Private Branch Exchange* (PBX), and the like.

Regardless of their type, all these switches have the same basic function—to route call traffic and to concentrate subscriber line traffic. Some of the more common switch concepts and designs are briefly described below along with a few example network applications.

3.4 Switching Functions

There are numerous functions of the switch within the network. These can be categorized into three basic groups: the elementary functions, the advanced functions, and the intermediate functions (Table 3.2).

Elementary switch functions include the process of connecting individual input and output line circuits (trunks) within the switch itself and the ability to control the distribution of communication traffic across clustered groups of line circuits (trunk groups). The ability to interconnect individual line circuits allows the transfer of voice or data signals between end subscriber units or between network nodes to take place in a controlled and selective manner. The switch's ability to direct or route traffic between individual line circuits, based on larger defined groups of line circuits, allows for more efficient and reliable control of large volumes of system traffic.

The more advanced functions are digit analysis, generation of call records, route selection, and fault detection. Digit analysis is the process of receiving the digits dialed by the customer, analyzing them, and determining what action the switch should perform based on this information, i.e., attempt to place a call to another party, connect them to an

TABLE 3.2 Switching Functions

Switching types	Functions
Elementary switching	Interconnection of input and output line circuits
	Control of communication traffic across line circuit groups
Advanced switching	Digit analysis
	Call record generation
	Route selection
	Fault detection
Intermediate switching	Monitor subscriber line circuits

operator for calling assistance, provide a recorded announcement stating that the digits dialed were in error, and the like. The process of generating a record or multiple records for any calling activity taking place within the switch is crucial to creating the corresponding billing records for these calls that, in the end process, result in the final bill being completed for the customer. Therefore, it is important that the switch produce an accurate account (record) of all call processing activity it performs.

The route selection function directs all system traffic within the switch to a specific set of facilities (transmission circuit, service circuit, and the like) based on routing tables developed and maintained by the equipment vendor and the system operator. Finally, the detection of errors or problems occurring within the switch's own hardware and software plus the identification of failures with any of the interconnected facilities is a required function of the switch to ensure the operating quality of the network.

An example of an intermediate switch function would be the monitoring of subscriber lines. This function involves the completion of regularly scheduled checks of all line circuits interconnected to the switch for proper operation, i.e., whether the circuit is still functional and able to carry system traffic. If a circuit is not performing properly then it will be taken out of service and the fault detection function notified for alerting the operations staff.

3.5   Circuit Switches

Telephone networks use circuit switches for the processing and routing of subscriber calls. Circuit switching can be either the space-division type or the time-division type or a combination of these two designs. These designs are used in the matrix of the circuit switch. The matrix is where the actual switching of line circuits or trunks takes place.

3.5.1   Space-division switching

In space-division switches the message paths are separated by space within the matrix. Thus, the name is derived. In Fig. 3.1 a simple space-division matrix is shown. At each end of the matrix are the wires (actual subscriber lines, line circuits, trunks, and the like) available for switching. This is represented as 1-N input lines and 1-N output lines. In this example the input j is connected to output k by closing the crosspoint (relay, contact, semiconductor gate, and the like) (j,k). The concept here is that only one row of input can be connected to a column of output.

In wireless networks using space-switches, each call has its own physical path through the network.

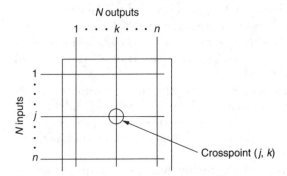

Figure 3.1   An $N \times N$ space-division matrix.

3.5.2   Time-division switching

In time-division switches the message paths are separated in time, hence the name. A simple diagram of a time-division switch is shown in Fig. 3.2 and will serve as a simple means to explain this switching concept. In Fig. 3.2, subscriber units $J1$ through Jn are in conversation with subscriber units $K1$ through Kn by means of a time-division switch. The actual input (originating subscriber) and output (terminating subscriber) line circuits are opened and closed by individual switching devices and indicated as such as $A1$ through An and $B1$ through Bn.

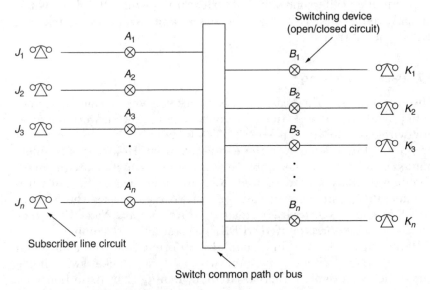

Figure 3.2   Time-division switch example.

(The use of the input and output line circuits does not matter in this explanation. It is used to show some similarity with the space-division switch matrix example.) In the time-division matrix the connection of subscribers takes place by controlling the operation of the selected switching devices, *A1* to *An* and *B1* to *Bn*.

For digital time-division switches to operate, the incoming transmitted voice signals for every phone call must be in a digitized and encoded format. A T-carrier transmission system can provide the proper format to allow direct interconnection to a digital time-division switch without any additional conversion equipment.

In summary, space-division switching involves the switching of actual circuit interconnections while time-division switching involves the switching of actual digitized voice samples within the switch matrix.

3.6 Circuit Switching Hierarchy

Within the North American PSTN, sometimes referred to as the *landline telephone system*, there are five classes of switches. These classes can be divided into the central office class 5 level switch and the remaining tandem-type switches of classes 1 through 4. The basic difference between these two categories of switches is that a class 5 local switch (also referred to as an end office) provides the ability to directly interconnect or interface to a subscriber's terminal equipment while a tandem-type switch will only provide interconnection to other switching equipment or systems. Thus, the local class 5 switch has the ability to terminate a call to one of its subscriber units while a tandem switch can only route calls to other destined nodes and never act as a final call delivery point (Fig. 3.3).

3.7 Packet Switching

The main function of a Packet Core Network in a wireless mobile system is to provide the overall IP connectivity. The packet network for fixed and wireless mobility is post RAN. The RAN is integral to the overall success and functionality of the wireless system. However, once information is in IP format then only the source and destination are required, enabling the packet network to be RAN technology agnostic. For wireless mobility systems, a packet core network comprises the *Serving GPRS Support Node* (SGSN)/*Gateway GPRS Support Node* (GGSN) or the *Packet Data Serving Node* (PDSN) packet network complex.

Packet networks are fundamentally different from the traditional voice, or circuit-switched communication systems. Packet switching is not connection oriented whereas circuit switching is, by definition, connection oriented.

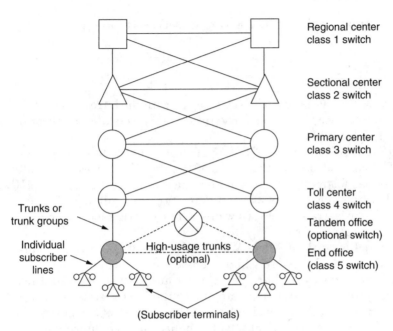

Figure 3.3 North American telephone switching hierarchy.

Traditional voice communication systems use circuit switches to provide the switching function of voice line circuits. In data communication systems, packet switches are used to perform the switching of data packets between the various nodes and computers in the network. Unlike the longer duration calls that require circuit switching in a telephone network, packet switching is better suited for the short burst-like transmissions of the data network. Packet switching involves the sorting of data packets from a single line circuit and switching them to other circuits within the network.

The packet network contains the functional network elements that work together for internetworking gateways to external networks such as public data network, corporate enterprise Intranet, and RAN. The core networks also provide the interface to network management and connect RANs with mobility management, security, and the like. The core network is designed to provide the operator with the ability to enhance the system offerings, commission features from third parties, and reduce overall operating costs.

With IP functionality at the mobile handset, applications like VoIP and Push-to-Talk functionality are becoming more widespread. The RAN and the selection of the proper handset are essential for the implementation of IP services; however it is the packet network that resides after the

RAN that treats and delivers the various services that are IP-based to the end user.

3.8 IP Networks

The IP network has gained wide-scale usage with the introduction of the graphical interface for browsing the Internet—a true killer application or rather killer enabler.

The IP network can be a LAN, WAN, Intranet, or the Internet to mention a few possibilities. The vastness of what an IP network can really comprise has fostered some of the misunderstanding. Basically, an IP network is another protocol which is another enabler allowing for more information to be transported. How and where it is transported can, and does, take on many forms.

Associated with the LAN or WAN is the use of an Intranet representing an internal network where members of the same campus, corporation, or whatever share resources—at least some—but these resources are not shared with anyone outside the group of users. Typically, the LAN/WAN involves connecting a series of computers to a hub that in turn might or might not be connected to an internal server, whether used for file sharing, database, web, or all the above.

There are a number of design books written with regard to IP networks and each has its own specific slant—service provider or vendor driven. Web sites like www.cisco.com are excellent sources for information related to IP design and questions on routing and IP address designs.

However, the heart of an IP network is the fostering of an ASP since providing a large pipe by itself will not result in additional revenue over the customer's life cycle.

The biggest issue with wireless data is that the network engineer needs to ensure that there is sufficient bandwidth to support the various applications and services offered and the correct CPU processing and memory to support the required additional services.

The delivery of IP data and the associated benefits of this exciting transport method are available with all 2.5G and 3G RAN as well as 802.11, 802.16, and 802.20 systems. The key difference is how they deliver the IP data defined by its latency and throughput.

In addition, the choice of using static or dynamic IP addressing schemes needs to be made. The use of static or dynamic IP addresses can be made with both public and private IP addresses. Static address is where the users are assigned the address and effectively own the address forever, or until they stop paying their bills. Dynamic DHCP involves assigning an address to the user for a limited time, referred to as a lease, and once the time expires the IP address assigned is reinserted into the pool for use by someone else. With DHCP the users do have a different IP address at different times when they use the system that is either public or private.

The use of static public IP addresses, while preferred by the end user, is not an efficient use of this scarce resource. Therefore, dynamic is the preferred method of IP address. In addition, private IP addresses should be used instead of public IP addresses.

3.9 IP Addressing

The issue of IP addressing is important to understand in any wireless data system design. No matter what the RAN or infrastructure vendor is, the use of IP addresses is essential to the systems operation. It is imperative that the IP addresses used for the network be approached from the initial design phase to ensure a uniform growth that is logical and easy to maintain over the life cycle of the system.

The use of IPv4 formatting is shown below. IPv6 or IPng is the next generation and allows for QOS functionality to be incorporated into the IP offering. However, the discussion will focus on IPv4 since it is the protocol today and has legacy transparency for IPv6.

Every device that wants to communicate using IP needs to have an IP address associated with it. The addresses used for IP communication have the following general format.

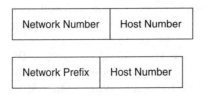

There are, of course, public and private IP addresses. The public IP addresses enable devices to communicate using the Internet while private addresses are used for communication in a LAN/WAN Intranet environment. A wireless data system can use both public and private IP addresses in order to facilitate the implementation of all its nodes.

Tables 3.3 and 3.4 represent the valid range of public and private IP addresses that can be used. The private addresses will not be recognized on the public Internet system and that is why they are used. Also, it is possible—and should be done—that a wireless data system can reuse private addresses within sections of its network, profound as this may sound. The concept is that the system can be segregated and the segregation allows for the reusing of private IP addresses ensuring a large supply of a seemingly limited resource. This is applicable when connecting various mobile data networks together.

TABLE 3.3 Public IP Addresses

Network address class	Range
A (/8 prefix)	1.xxx.xxx.xxx thru 126.xxx.xxx.xxx
B (/16 prefix)	128.0.xxx.xxx thru 191.255.xxx.xxx
Class C (/24 prefix)	192.0.0.xxx thru 223.255.255.xxx

The public addresses are broken down into A, B, and C addresses with their ranges as shown in Table 3.3.

The private addresses that should be used are shown in Table 3.4.

To facilitate the use of IP addressing, the use of subnetting further helps refine the addressing by extending the effective range of the IP address itself. The IP address and its master subnet directly affects the number of subordinate subnets that can exist and from those subnets the number of hosts—users—that can be assigned to that subnet. Table 3.5 shows the number of users that a subnet can support.

It is important to note that the IP addresses assigned to a particular subnet include not only the host IP addresses but also the network and broadcast address. For example, the 255.255.255.252 subnet that has two hosts requires a total of four IP addresses to be allocated to the subnet—two for the hosts, one for the network, and the other for the broadcast address. Obviously, as the number of hosts increases with a valid subnet range, the use of IP addresses becomes more efficient. For instance, the 255.255.255.192 subnet allows for 62 hosts and uses a total of 64 IP addresses.

Therefore, you might say why not use the 255.255.255.255.192 subnet for everything. This, however, would not be efficient either and so an IP address plan needs to be worked out in advance since it is extremely difficult to change once the system is being or has been implemented.

What is the procedure for defining IP addresses and subnetting? The following rules apply when developing the IP plan for the system. The same rules are used for any LAN or ISP that is designed. There are four basic questions that help define the requirements and they are as follows:

TABLE 3.4 Private IP Addresses

Private network address	Range
10/8 prefix	10.0.0.0 thru 10.255.255.255
172.16/16 prefix	172.16.0.0 thru 172.31.255.255
192.168/16 prefix	192.168.0.0 thru 192.168.255.255

TABLE 3.5 Subnets

Mask	Effective subnets	Effective hosts
255.255.255.192	2	62
255.255.255.224	6	30
255.255.255.240	14	14
255.255.255.248	30	6
255.255.255.252	62	2

1. How many subnets are needed at present?

2. How many are needed in the future?

3. What is the number of users or APs on the largest subnet at present?

4. What is the number of users or APs on the largest subnet in the future?

Therefore, using the above method an IP plan can be formulated. It is important to note that the IP plan should not only factor in the design the end customer needs but also in the one the wireless data provider needs.

Specifically, the wireless operator's needs will involve IP addresses for the following platforms as a minimum. The platforms requiring IP addresses are constantly growing as more and more functionalities for the devices are added through SNMP.

- Base stations
- Radio elements
- Microwave Point to Point
- Access points
- Customer terminals (i.e., mobile phones)
- Routers
- ATM switches
- Workstations
- Servers

The list can and will grow when you tally all the devices within the network both from a hardware and network management aspect. Many of the devices listed above require multiple IP addresses in order to ensure their functionality of providing connectivity from point A to point B. It is extremely important that the plan follows a logical method.

A suggested methodology is to

1. List out all the major components that are, will be, or could be used in the network over a 5- to 10-year period.

2. Determine the maximum number of these devices that could be added to the system over 5 to 10 years. A suggestion is to do this calculation by BSC or BTS.

3. Determine the maximum number of packet data devices per sector that can be added.

4. Determine the maximum number of customers who can be connected to each of the packet data devices (could be more than one).

5. Determine the maximum number of physical sectors that could be used per base station.

6. Determine the maximum number of AP radios and their type for each sector and base station.

Keep in mind that if there are multiple customers per data terminal, the number of IP addresses required will in all likelihood increase depending on the type of service offered.

For instance, the possibility of using public addresses for all the devices is not practical, but some public addresses will be required. Therefore, the use of private IP addresses will be required to ensure that all the required devices have an associated IP address that has the correct network and subnet associated with the particular service.

Depending on how the network is laid out, the reuse of IP addresses can take place if each city or region had its own domain name and of course *Network Address Translation* (NAT). The objective is to keep as many devices on the same network as possible so that the number of retranslations and hops is kept to the minimum.

For discussion purposes the IP assignment scheme used for access points should be defined prior to the initial deployment. The IP assignment process could be designed so that the IP addresses are associated with a particular grid number instead of a sector of a particular cell. However, the IP address by sector will probably prove more user friendly for operations than having a grid and then another lookup cross-reference table to the cell and sector.

Naturally, your particular requirements will be different and the IP address method implemented. The concept is to make the plan and then work the plan.

3.10 Soft-Switch

Soft-switches are making their way into the wireless mobile networks. Soft-switches have several significant advantages over traditional legacy circuit switches.

But what is a soft-switch?

There is no single answer to the question posed because there is little or no clear definition of what a soft-switch is, unlike descriptions for traditional switches like class 5 or class 4 switches, to mention a few. A soft-switch is a distributed architecture that is packet based. The soft-switch distributed switching architecture is the convergence of traditional voice and packet (IP) services.

The distributed architecture is possible largely because it is a packet-based network and therefore the various elements—BSCs and servers—do not need to be centrally located, as is done traditionally. Conventionally, at the heart of a wireless mobile system is the class 5 switch that performs all the traditional voice-centric service treatments and is a large concentration node.

Soft-switches also interface with traditional circuit-switched services through standard interfaces providing TDM functionality. Therefore a soft-switch can support both Class 4 and Class 5 service applications and deliver traditional PSTN, *Integrated Services Digital Network* (ISDN), VoIP/FoIP, and 3G services, over both wireline and wireless networks.

Soft-switching enables the wireless operator to distribute the call-processing functionality thereby enhancing disaster recovery and removing single points of failure like the traditional class 5 switch in a mobile network. The soft-switch also enables the operator to better load balance the system and introduce new features and functionality in a more efficient way. This ability is due to the inherent structure of a soft-switch architecture where the capacity is very scalable and can be grown in a modular and incremental fashion.

With any new or emerging technology there are numerous variations or types of soft-switches that are available. Pressure continues for a wireless operator to exploit packet data offerings. Therefore interoperability becomes paramount as different communications networks begin to converge. With increased processing power and memory for computing, the soft-switches rely on ATM and IP as their fundamental platform.

Soft-switching is based on open interfaces allowing for standard off-the-shelf equipment, such as commercial computing platforms and industry standard network components, to be used enabling a mix-and-match solution that can, if done properly, significantly lower the total cost of operation as compared to the traditional proprietary closed solutions. A soft-switch requires less room, less power to operate, and can be deployed much more quickly than traditional legacy circuit switches further improving the flexibility and usefulness of this architecture.

In addition, soft-switch architectures allow rapid feature development and high-core networking flexibility. Both circuit and packet networks are converging, requiring both voice and data support with minimal impact on the network or its components. This will create new vertical and geographic markets that take advantage of new trends for economic viability.

Convergence to a single network will merge today's separate voice and data networks into multipurpose packet networks supporting voice, data, and multimedia. Operating, administering, and maintaining one network significantly reduces maintenance costs for telephone operators. This unique solution eliminates or reduces the need to backhaul voice/data traffic for local calls and reduces operational costs. Open-standard technologies allow for greater flexibility.

Packet technology greatly reduces the bandwidth needed to support a voice call. This means that the additional call capacity is increased with existing bandwidth, thereby decreasing the need to build-out these telecommunications facilities. Point-to-point routing eliminates routing through an intermediate centralized switch. With soft-switches, wireless operators can continue to offer and deploy traditional wireless voice services.

But most important, the deployment of soft-switches is a clear method to "futureproof" the core network. This is because soft-switch technology brings all the benefits of 3G next generation networks to current 2G and 2.5G using remote intelligent media gateways to adapt standard base stations to the 3G and 2.5G core network.

3.11 ATM

Asynchronous Transfer Mode transport has been used by the PSTN for some time now but with the introduction of mobile data and the need to transport data and possibly voice, the desire to use a transport medium that can handle both continuous traffic and bursty traffic is essential. With the continuous push from more IP-related services, the network designer for a mobile data system is faced with the dilemma of ensuring different grades of service, QOS, while at the same time not overdimensioning the various pipes within the network so as to reduce recurring facility costs.

Entire books are written on the design and functionality of ATM switching. I do not intend to rewrite those excellent books that can be found in the reference portion of this chapter. Instead, the objective is to show how the ATM platforms fit into the wireless mobile service platforms, depending of course on the services and functions that are offered by the system.

ATM is, by default, a connection-oriented transport method that is excellent for transporting all types of information content—specifically data traffic—and can also transport voice services effectively. The ATM switch performs two major functions, routing (self and label) and header translation.

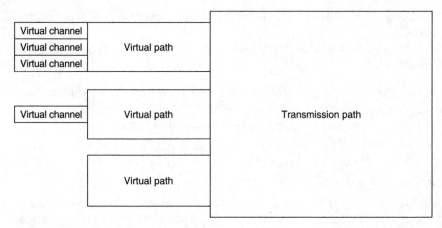

Figure 3.4 Virtual path.

The ATM routing types are: *sequential routing* (SR), *random alternative routing* (RAR), *least loaded routing* (LLR), and *minimum cost routing* (MCR).

In an ATM switch the connections are referred to as *Virtual Circuits* (VCs) but the VC is just a container for tributary paths, called *virtual paths* (VPs). There can be multiple VPs within a Virtual Circuit as illustrated in Fig. 3.4.

ATM is able to provide greater throughput in the network because it uses fixed-length packets called cells. A cell consists of 53 bytes of which 48 bytes are for data and 5 bytes for the header. Figure 3.5 illustrates the

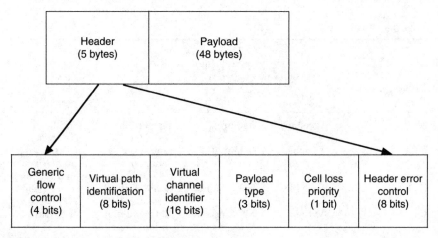

Figure 3.5 ATM cell format.

makeup of an ATM cell and the various components that comprise the ATM cell itself.

Referring to Fig. 3.5, the various fields have specific functions. More specifically:

- *Generic Flow Control (GFC).* This field's purpose, as is evident by its name, is to control the traffic flow on ATM connections. It is also used for CBR services for the purpose of controlling jitter.

- *Virtual Path Identifier (VPI).* The basic function of this field is to identify the VPI used for connecting the Virtual Channel Identifier (VCI).

- *Virtual Channel Identifier (VCI).* It identifies the channel.

- *Payload Type (PT).* It provides the type of data contained within the cell. Types of information the PT will identify include user data, signaling data, congestion information, and maintenance data.

- *Cell Loss Priority (CLP).* This is the field that identifies if the cell should be discarded during congestion.

- *Header Error Control (HEC).* This field's purpose is to provide the error correction.

The ATM transport method allows for different service classes and these classes are directly related to the *ATM Adaptation Layer* (AAL). There are of course 5 AALs and each is meant to transport a particular type of service. The type of service for each of the AALs and the connection type are listed in Table 3.6 for easy reference.

Table 3.7 is interesting but without relating the specific AAL levels to possible service offerings, it has little value. Therefore, Table 3.7 is meant to help further refine Table 3.6.

TABLE 3.6 AAL

Service Class	AAL	Bit rate	Timing relationship between source and destination	Connection type	Applications
A (1)	1	Constant	Required	Connection oriented	CES CBR video
B (2)	2	Variable	Required	Connection oriented	Rt-VBR audio/video (multimedia)
C (3)	5 (3)	Variable	Not required	Connection oriented	Frame Relay FTP
D (4)	5 (4)	Variable	Not required	Not connection oriented (i.e., connectionless)	IP, SMDS

TABLE 3.7 Six ATM Class/Type

Class/type	Bandwidth guarantee	Real-time traffic	Bursty traffic	Congestion feedback
Constant bit rate (CBR)	Yes	Yes	No	No
RT-VBR Variable Bit Rate, Real Time	Yes	Yes	No	No
Variable bit rate, nonreal time (NRT-VBR)	Yes	No	Yes	No
Available bit rate (ABR)_	Yes/No	No	Yes	Yes
Unspecified bit rate (UBR)	No	No	Yes	No

- *Constant bit rate (CBR).* This supports applications that require a fixed data rate being always provided. Examples of where CBR is applied involve TDM circuits, voice and leased lines like T1/E1.

- *Real-time variable bit rate (rt-VBR).* This service supports applications requiring real-time data flow control. Examples of where rt-VBR is used include video conferencing. The main difference between CBR and rt-VBR is that rt-VBR has tighter timing controls than used for CBR.

- *Nonreal time variable bit rate (nrt-VBR).* This service supports applications that, as the name suggests, do not require exacting timing between source and destination. Some applications for nrt-VBR involve e-mail and multimedia applications.

- *Available bit rate (ABR).* This service is really an enhancement to UBR in that a minimum and peak cell rate can be defined. ABR has priority over UBR traffic. Effectively, ABR is a managed best-effort service.

- *Unspecified bit rate (UBR).* Also referred to as best-effort service, this service supports applications that can tolerate variable delays between source and destination plus the possibility of cell loss. UBR traffic uses the bandwidth that is left over after CBR and VBR services have taken their share of the pipe. The UBR service allows more utilization of the ATM network by passing UBR traffic at different rates between CBR and VBR allocations.

3.11.1 ATM networks

What is an ATM network? Well, to start with an ATM network can consist of a single ATM switch or multiple ATM switches. A key concept to keep in mind is that when you want to leave the ATM media world it will require converting to the appropriate protocol like TDM or IP, which is usually done with an ATM edge switch.

ATM switches are being deployed in base stations as a method of concentrating TDM and IP traffic onto a single pipe for backhaul transport. The issue that needs to be considered by the network engineer is the introduction of ATM switching platforms that are effectively protocol converters that increase operating expenses.

With that said, ATM systems support two general types of interfaces:

UNI. User Network Interfaces

NNI. Network Network Interface

Figure 3.6 represents the location of the different interfaces. Basically, the private NNI occurs when connecting to ATM switches within your own network and the public UNI occurs when you connect to the public ATM switch or cloud. The reason this is mentioned is that PNNI signaling is not available with a UNI interface.

There are also several types of NNIs for ATM networks:

PNNI. Private Network-to-Network Interface

B-ICI. Broadband Intercarrier Interface

B-ISUP. Broadband ISDN Services User Part

IISP. Interim Interswitch Signaling Protocol

ATM routing is centered around circuits of two types—virtual paths and virtual connections (VCI). Included is the use of SVCs and PVCs.

Switched Virtual Connection (SVC). It is a connection that is set up dynamically based on the need for the connection. Specifically, every

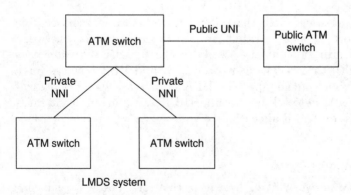

Figure 3.6 ATM network interfaces.

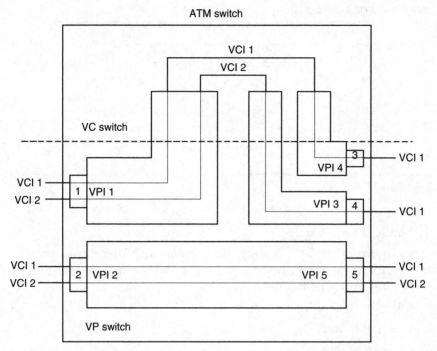

Figure 3.7 ATM switch.

time service is requested, the path taken from the source to the destination can change based on resources available. SVCs make best use of the network's facilities by increasing its utilization.

Permanent Virtual Connection (PVC). It is a predefined route that is programmed into the ATM switch via a craft person. In Fig. 3.7 all of the paths could be considered PVCs as long as the VPI/VCI and ports remain the same.

Table 3.8 is an illustration of the routing functions that are associated with an ATM switch. Naturally, there is more than the simple port mapping illustrated hereafter, but the diagram helps facilitate the issue of VPI and VCI mapping.

3.11.2 ATM design aspects
The design aspects for ATM switch designs focus on several key attributes:

TABLE 3.8 ATM Switch Mapping

Input			Output		
Port	VPI	VCI	Port	VPI	VCI
1	1	1	3	4	1
1	1	2	4	3	1
2	2	1	5	5	1
2	2	2	5	5	2
5	5	1	2	2	2
5	5	2	2	2	2
3	4	1	1	1	1
4	3	1	1	1	2

1. Traffic characteristics

   - *Burstiness.* It is commonly used to measure how infrequent the traffic volume and rate are between the source and destination. Burstiness = peak rate/average rate
   - Traffic delay tolerance
   - Response time
   - Capacity throughput

2. Cell delay

3. Cell loss

4. Congestion

5. Ports available

The first topic traffic characteristics, is important to factor in since an ATM switch is designed to handle CBR, VBR, and UBR traffic. The type of traffic, QOS associated with each, and of course the burstiness of the traffic all factor into the overall throughput for the switch. Typical design limits put the desired throughput for the design to be 70 percent of the allowable limit of the ATM switching platform. Throughput is not the same as the number of ports.

The other topics like cell delay factor in the ability to deliver a particular time-sensitive traffic. The cell delay is important for the individual switch but is in reality related to the overall network design and in particular to the path the cell has to transverse from start to finish.

Cell loss is important for many protocols that do not have error correction. Additionally, if there is a lot of cell loss, the link loads increase due to retransmissions and this has a direct effect on the throughput and cell-delay factors.

Congestion by itself is an important issue to avoid for an ATM switch design. During high-congestion periods, heavy cell loss can occur—by design—resulting in more congestion due to the amount of retransmission that takes place.

The port issue is important when connecting to different ATM networks and providing circuit emulation to a TDM platform. Many times the number of ports is the driving issue for ATM edge switches and not the other design parameters. The ports for an ATM switch can run at 100 percent utilization. However, in the initial design phase with unknown traffic projections, the desired level is 70 percent of the estimated load for any ATM switch platform chosen at the design end point. This is of course based on the premise that the ATM platform chosen has the available card slots—pods—to accommodate this potential growth.

3.12 Facility Sizes

Table 3.9 is a brief chart showing the different facilities that a wireless data operator may have to interface to and the associated bandwidth of each. While Table 3.10 is a quick reference regarding the similarities between an SONET and an SDH fiber system.

3.13 Demand Estimation

The demand that the system will need to meet (transport) and the associated bandwidth needed can be reasonably determined given historical data intermixed with marketing forecast data. The marketing forecast data are important for both a new and an existing system since this is the best way to estimate the amount of traffic that a system will need to transport, and of course the media type associated with the traffic load.

TABLE 3.9 Facility Sizes

Signal level	Carrier system	Number of DS1 systems	M bits/sec
DS0	DS0	1/24	0.064
DS1	T1	1	1.544
DS1C	T1C	2	3.152
DS2	T2	4	6.312
DS3	T3	28	44.736
DS4	T4	168	274.76
OC1	OC1	28	51.84
OC3	OC3	84	155.52
OC12	OC12	336	622.08
OC48	OC48	1344	2488.32

TABLE 3.10 Fiber Sizes

Signal type	Mbps	SDH
STS-1/OC1	51.84	-
STS-3/OC-3	155.52	STM-1
STS-12/OC-12	622.08	STM-4
STS-24/OC-24	1244.16	STM-8
STS-48/OC-48	2488.32	STM-16

The key difference between a new and an existing system design lies in the issue of having a baseline from which to begin the forecast from. If the system or service is new then there is no baseline. However, if the system is in operation there should be some traffic being carried by the system and therefore the marketing forecast is an addition to the current traffic being carried or designed.

Design traffic = current carried traffic + forecasted traffic

The key points are trying to determine how far in advance the study needs to be done and the frequency of the study. The recommendation is that the study should not be done in detail for more than 2 years. The first year should be broken down by quarters while the second year should be done in 6 month intervals. The traffic estimation will need to be revisited on a 3 or 6 month basis for the life of the system to ensure proper dimensioning is taking place and to account for the hotspots or lack of take rates estimated.

For the forecasting, there are several key elements that the network engineer needs to factor in to the forecast and design.

1. Types of services

2. Volume of traffic for each service

3. Platform requirements and growth

4. Connectivity between base station or access point and central office or concentration node

5. Connectivity between concentration node and the various transport providers (i.e., PTT, CLEC, IP)

6. Time frame of study

7. Current traffic utilization

The list provided can, and should, be tailored to your specific requirements and needs.

3.14 VoIP

The advantages of VoIP have been gaining acceptance in recent years as the benefits of this transport method for voice are being understood. Overall, the transport of data via IP is widely used but transporting voice using IP, VoIP, is not as well known or understood. VoIP has been used for many years and has seen maximum usage as a backbone, core, and transport mechanism and not focused on the end or edge user or application.

VoIP is an exciting application that is starting to be exploited by mobile and fixed wireless service providers. However, the proper treatment or discussion regarding the details of VoIP are best found in the references included at the end of this chapter.

The original standards activity for VoIp was defined in H.323 that has the name *Packet-Based Multimedia Communication Systems*. This standards-wide use was a direct result of offering it as freeware by Microsoft. There is, however, an alternative standard which is currently in competition with H.323 and that is *Media Gateway Control Protocol* (MGCP), also called *Single Gatway Control Protocol* (SGCP) and *Session Initiation Protocol* (SIP).

IP has a number of advantages over traditional circuit-switching. The most notable of these is the fact that it can leverage today's advanced voice coding techniques, such as the *Adaptive Multirate* (AMR) coder used in *Enhanced Data Rates for Global Evolution* (EDGE) and *Universal Mobile Telecommunications Service* (UMTS) networks, besides the other 2.5G/3G systems. Thus, voice can be transported with far less bandwidth than the 64 kbps used in traditional circuit-switched networks.

IP is a layer 3 protocol in the *Open Systems Interconnection* (OSI) seven layer protocol stack. IP by itself is inherently unreliable and provides no protection against loss of packets or delays. To combat this *Transmission Control Protocol* (TCP) is used to ensure error-free in-sequence delivery of packets to the destination application. This protocol resides on the layer above IP.

When a VoIP session is to be set up, the session or application data are first passed to TCP (where a TCP header is applied), then passed to IP (where an IP header is applied), and then forwarded through the network. The information contained in the TCP header includes—among other things—source and destination port numbers that allow for the identification of the applications at each end, sequence numbers and acknowledgement numbers that allow for detection of lost packets, and a checksum, which allows for detection of corrupted packets.

At the heart of VoIP is the desire to provide good speech quality. Speech quality is essential and traditional voice coding uses 64 kbps of bandwidth, G7.11, which is circuit-switched-based. Alternatives to the

G.711 coding scheme are seen in mobile wireless systems through the use of vocoders that use advanced coding scheme to emulate or achieve the quality of G.711. Other requirements include low transmission delay, low jitter (delay variation), and the requirement that everything transmitted at one end is received at the other (i.e., low loss).

In order to minimize delay *User Datagram Protocol* (UDP), layer 4, is used. UDP, however, offers no protection against packet loss. Given the choice between UDP and TCP, the issue is whether we consider minimizing delay to be more important than eliminating packet loss. The answer is that, for speech, excessive delay and excessive jitter are far more disturbing than occasional packet loss. Consequently, when transporting voice, UDP is chosen at layer 4 rather than TCP.

It is clear, however, that something more than UDP is required if VoIP is to offer reasonable voice quality. In order to fulfill these needs, a protocol known as the *Real-Time Transport Protocol* (RTP) was developed. This protocol resides above UDP in the protocol stack. Whenever a packet of coded voice is to be sent, it is sent as the payload of an RTP packet. That packet contains an RTP header, which provides information such as the voice coding scheme being used, a sequence number, a timestamp for the instant at which the voice packet was sampled and an identification for the source of the voice packet.

RTP has a companion protocol, the *RTP Control Protocol* (RTCP). RTCP does not carry coded voice packets. Rather, RTCP is a signaling protocol that includes a number of messages, which are exchanged between session users. These messages provide feedback regarding the quality of the session.

However, RTP and RTCP do not guarantee minimal delay, low jitter, or low packet loss. In order to do that, other protocols are required. RTP and RTCP simply provide information to the applications at either end so that those applications can deal with loss, delay, or jitter with the least possible impact on the user.

If the VoIP session stays within a data network then the use of traditional landline, class 5 switches, are not required. But in many instances the call could originate as a VoIP call but terminate as a circuit-switched call at a residential or business phone that does not use VoIP. In order to achieve communication between the IP client and the traditional circuit-switched user a media gateway is required, which is essentially a protocol converter that ensures that speech quality is achieved at the interface.

3.15   OSI Levels

The OSI levels are important to understand when converging technologies and access platforms commingle. The OSI layers consist of seven layers, 1 through 7, as shown in Table 3.11. Referring to Table 3.11, layer

TABLE 3.11 OSI Layers

OSI layer	Layer name	Comments
7	Application	Used for connecting the application program or file to a communications protocol
6	Presentation	Performs the encoding and decoding functions
5	Session	Establishes and maintains the connection for the communication processes in the lower OSI layers
4	Transport	Error correction and transport, both Tx and Rx are performed here
3	Network	Switching and routing functions for the MSC are done here
2	Data link	Receives and sends data over the physical layer
1	Physical	Actual media used for sending and receiving the communications. Radio, or fiber optic wires are two examples

7 is the highest OSI layer while layer 1 is the lowest. OSI layer 1 is the physical layer and is the starting point for the OSI layer hierarchy.

References

Azzam, A. A., *High-Speed Cable Modems*, McGraw-Hill, New York, 1997.
Bates, R. J., and D. Gregory,*Voice and Data Communications Handbook*, Signature edition, McGraw-Hill, New York, 1998.
Black, U., *TCP/IP and Related Protocols*, McGraw-Hill, New York, 1992.
Collins, D., *Carrier Grade Voice Over IP*, McGraw-Hill, New York, 2001.
Dudendorf, V., *Wireless Data Technologies*, Wiley, England, 2003.
Goralski, W., *ADSL and DSL Technologies*, McGraw-Hill, New York, 1998.
Guizani, M., and A. Rayes, *Designing ATM Switching Networks*, McGraw-Hill, New York, 1999.
McDysan, D., and D. Spohn, *ATM Theory and Applications,* Signature edition, McGraw-Hill, New York, 1999.
Ohrtman, F., and K. Roeder, *Wi-Fi Handbook*, McGraw-Hill, New York, 2003.
Pahlavan, K., and P. Krishnamurthy, *Principles of Wireless Networks*, Prentice Hall, New Jersey, 2002.
Russell, T., *Signaling System Number 7*, 2d ed., McGraw-Hill, New York, 1998.
Smith, C., *Wireless Telecom FAQ*, McGraw-Hill, New York, 2000.
Smith, C., *LMDS*, McGraw-Hill, New York, 2000.
Smith, C., and C. Gervelis, *Cellular System Design and Optimization*, McGraw-Hill, New York, 1996.
Winch, R., *Telecommunication Transmission Systems*, 2d ed., McGraw-Hill, New York, 1998.

Mobile Wireless Systems

There is a vast sea of mobile wireless systems that are in existence today with more being proposed or being implemented. Some of the wireless systems are derived from standards and others from the introduction of proprietary protocols and configurations. However, each particular wireless technology has its own unique advantages and disadvantages that are necessary to understand when determining enhancements and service offerings that they can properly support. Ultimately, the wireless mobility system or systems that are deployed in the network need to meet the current business objectives and be future proof within reason.

The main Mobile Wireless Systems that will be discussed briefly are:

- CDMA2000
- GSM/GPRS/EDGE
- UMTS
- 802.20 (covered in Chap. 7)

There are also supplemental or adjunct wireless systems that also enhance the existing wireless offering. The enhanced wireless systems that augment the mobility systems are:

- 802.11 b
- 802.11 g
- 802.16 a
- 802.16 e
- 802.15

The 802.11 and 802.16 platforms are discussed in Chaps. 5 and 6 respectively.

For this chapter, a brief description of IMT-2000 along with several of the most relevant mobility technology platforms will be covered. More explicit detail for each of the mobility platforms can be obtained from the references that are at the end of the chapter.

4.1 IMT-2000

IMT-2000 is the ITU specification *International Mobile Telecommunications 2000*. IMT-2000 is more commonly called 3G. The term 3G has received and continues to receive much attention as the enabler for high-speed data for wireless mobility.

IMT-2000 is a radio and network access specification defining several methods or technology platforms that meet the overall goals of the specification. The IMT-2000 specification is meant to be a unifying specification allowing for mobile and some fixed high-speed data services using one or several radio channels coupled with fixed network platforms for delivering the services envisioned.

- For a service to claim to be 3G, it must meet the IMT-2000 standard, or rather specification, that includes Global standard
- Compatibility of service within IMT-2000 and other fixed networks
- High quality
- Worldwide common frequency band
- Small terminals for worldwide use
- Worldwide roaming capability
- Multimedia application services and terminals
- Improved spectrum efficiency
- Flexibility for evolution to the next generation of wireless systems
- High-speed packet data rates
  - 2 Mbps for fixed environment
  - 384 Mbps for pedestrian
  - 144 kbps for vehicular traffic

Some of the more salient issues for 3G involve the interoperability and data rates. However, there is no single platform that makes up IMT-2000. In fact, IMT-2000 is a collection of specifications that are designed to have some level of interoperability. Figure 4.1 shows the relationship between the various wireless mobility platforms that comprise IMT-2000.

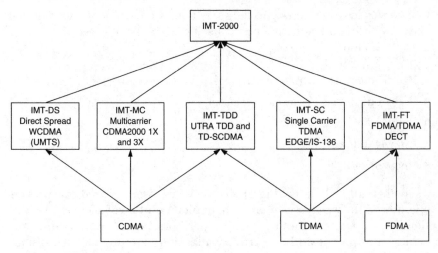

Figure 4.1   IMT-2000.

True interoperability between technologies is as elusive as the wireless data killer application, but 802.20 promises to fill interoperability void. However, is not part of the IMT-2000 specification.

4.2 Wireless Mobility Platforms

There are a multitude of wireless mobility platforms that are in service today. Presently the wireless industry is in transition between 2G and 3G. There is an interim set of technology platforms referred to as 2.5G. The exact definition of 2G, 2.5G, and 3G is a constant source of debate. However, the following is a simple definition for each of the generations.

First Generation (1G) includes all of the analog technologies, primarily AMPS and TACS.

Second Generation (2G) includes GSM, IS-136 and IS-95, and iDEN.

Two and One-Half Generation (2.5G) includes GSM/GPRS/EDGE, CDMA2000 1xRTT and EVDO, iDEN.

Third Generation (3G) includes WCDMA, CDMA2000-EVDV, and SCDMA.

First Generation mobility systems were analog and proved to be a great advance in communication mobility. The success of mobile communication led to the need for improvements culminating in 2G systems. The introduction of 2G mobility systems while focused on voice transport brought about numerous improvement or enhancements for the mobile wireless operators and their customers. The major benefits associated with the introduction of a 2G system are listed here.

- Increased capacity over analog
- Reduced capital infrastructure costs
- Reduced capital per subscriber cost
- Reduced cellular fraud
- Improved features
- Encryption

The benefits when looking at the above list were geared toward the operator of the wireless system. The implementation of 2G was a reduction in operating costs for the mobile operators either through improved capital equipment and spectrum utilization to reduction in cellular fraud. The improved features were centered around *short message services* (SMS) that benefited the subscriber. The onslaught of 2G systems, however, benefited the customer in the primary aspect that the overall cost to the subscriber was significantly reduced.

Second Generation, as a generalization is used to describe the advent of digital mobile communication for cellular mobile systems. When cellular systems were being upgraded to 2G capability the description at that time was digital and there was little, if any, indication of 2G since voice was the service to deliver, not data. Personal Communication Systems at the time of their entrance were considered the next generation communication systems and boasted about new services that the subscriber would want, which could be readily provided by this new system or systems. However, PCS services took on the same look and feel as those originating from the cellular bands.

Digital or digital modulation is now prevalent throughout the entire wireless industry. Digital communication refers to any communication that uses a modulation format that relies on sending the information in any type of data format. More specifically, digital communication is where the sending location digitizes the voice communication and then modulates it. At the receiver the exact opposite is done.

Digital radio technology is deployed in a cellular/PCS/SMR system primarily to increase the quality and capacity of the wireless system over its analog counterpart. The use of digital modulation techniques enables the wireless system to transport more bits/Hz than would be possible with analog signaling using the same bandwidth. However, the service offering for 2G is a voice offering.

For the cellular 1G operators there were several options or rather decisions to make on how to integrate the new system into the existing analog network. The exception was Europe, which through regulation defined *Global System for Mobile Communications* (GSM) as the technology platform to use. In addition, whether the technology platform was defined by regulation or not, the PCS operators when entering the market did not have any legacy system problem because there was no legacy system.

The integration with the existing 1G legacy systems was therefore an issue that affected only the analog systems operating in the 800/900-MHz bands. The primary 2G technologies that were applicable involved GSM,TDMA, CDMA, and iDEN Radio Access Systems. Regardless of technology platform—radio access—chosen, the issues regarding 2G deployment were

- Capacity
- Spectrum utilization
- Infrastructure changes
- Subscriber unit upgrades
- Subscriber upgrade penetration rates

The fundamental binding issue with 2G is the use of digital radio technology for transporting the information content. It is important to note that while 2G systems used digital techniques to enhance their capacity over analog, their primary service was voice communication. When 2G systems were being deployed, 9.6 kbps was more than sufficient for existing data services, usually mobile fax. A separate mobile data system was deployed in the United States, called *Cellular Data Packet Data* (CDPD), which was supposed to meet the mobile data requirements. However, this too has become a system of the past. In essence, 2G systems were deployed to improve the voice traffic throughput over an existing analog system.

Digital radio technology was deployed in cellular systems using different modulation formats with an attempt to increase the quality and capacity of the existing cellular systems. As a quick point of reference in an analog cellular system, the voice communication is digitized within the cell site itself for transport over the fixed facilities to the *Mobile Telephone Switching Office* (MTSO)/*Mobile Switching Center* (MSC). The voice representation and information transfer used in 1G/*Advanced Mobile Phone System* (AMPS) cellular was analog and it is this part in the communication link that digital transition was focusing on.

The digital effort there is meant to take advantage of many features and techniques that are not obtainable for analog cellular communication. There are several competing digital techniques that have been deployed in the cellular arena. The digital techniques for cellular communication fall into two primary categories—AMPS and TACS spectrum. For markets employing the TACS spectrum allocation GSM was the digital modulation technique chosen. However, for AMPS markets the choice is between TDMA and CDMA radio access platforms. In addition to the AMPS/TACS spectrum decision there is the iDEN radio access platform in that it operates in the specialized mobile radio (SMR) band that is neither cellular nor PCS. With the introduction of the PCS licenses there are three fundamental competing technologies—CDMA, GSM, and TDMA. Which technology platform is best depends on the application desired and at present each platform has its pros and cons including whether there is a regulatory requirement to utilize one particular platform or not.

There were several options available for the 1G operators to follow and the level of complexity for decisions continued for operators when they used 2G and wished to migrate to 3G. The migration path or possible paths that a wireless mobility operator can use to go from 1G or 2G to 3G is shown in Fig. 4.2. The rationale behind discussing 1G and 2G

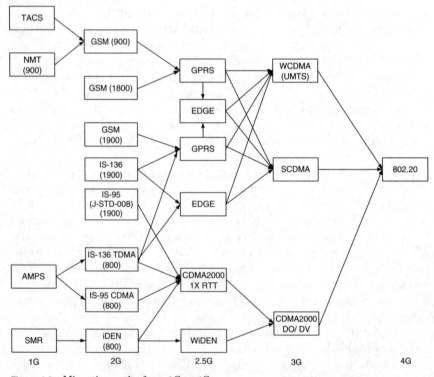

Figure 4.2   Migration paths from 1G to 4G.

TABLE 4.1   Wireless Technology Platforms: (a) Cellular/SMR (b) PCS

Cellular and SMR systems

	AMPS	IS-136	GSM	IS-95 (CDMA2000)	iDEN
Multiple Access Method	FDMA	TDMA/FDMA	TDMA	CDMA/FDMA	TDMA
Modulation	FM	Pi/4DPSK	Pi/4DPSK	QPSK	16 QAM
Radio Channel Spacing	30 kHz	30 kHz	30 kHz	1.25 MHz	200 kHz
Users/ Channel	1	3	3	64	3/6/2
Number Channels	832	832	600	9 (A) 10 (B)	600
CODEC	NA	ACELP/ VCELP	ACELP	CELP	RELP- LTP

PCS

	IS-136	CDMA 2000 IS-95	DCS1800 (GSM)	DCS1900 (GSM)
Base Tx MHz	1930–1990	1930–1990	1805–1880	1930–1990
Base Rx MHz	1850–1910	1850–1910	1710–1785	1850–1910
Multiple Access Method	TDMA/FDMA	CDMA/FDMA	TDMA/FDMA	TDMA/FDMA
Modulation	Pi/4DPSK	QPSK	0.3 GMSK	0.3 GMSK
Radio Channel Spacing	30 kHz	1.25 MHz	200 kHz	200 kHz
Users/channel	3	64	8	8
Number Channels	166/332/498	4–12	325	25/50/75
CODEC	ACELP/VCELP	CELP	RELP-LTP	RELP-LTP
Spectrum Allocation	10/20/30 Mhz	10/20/30 Mhz	150 MHz	10/20/30 Mhz

migration is the quest for delivering mobile wireless packet data services with 3G and beyond, to 4G.

Table 4.1 represents some of the different technology platforms in the cellular, SMR, and PCS bands.

What follow next are individual discussions regarding the major wireless technology platforms. Since some of the standards cross the boundaries between 2G, 2.5G, and soon to be 3G, it is felt that, for brevity, the best method to proceed here is to cover each technology platform separately. The focus will be on CDMA2000, GSM, and UMTS.

4.3 CDMA2000 (1xRTT, 1xEVDO, 1xEDV)

CDMA2000 comes under the specification of IS-2000, which is backward compatible with IS-95A and B plus J-STD-008 specifications, which collectively are called CDMAOne. CDMA2000, while being a 3G specification, is also backward compatible with CDMAOne systems, allowing operators to make strategic deployment decisions in a graceful fashion.

Since CDMA2000 is backward compatible with existing CDMAOne networks, the upgrade or rather changes to the network from a fixed network aspect can be done in stages. More specifically, the upgrades or changes to the network involve the BTS with Multimode Channel Element cards, the BSC with IP routing capability, and the introduction of the *Packet Server Data Network* (PSDN). The radio channel bandwidth is the same for CDMA2000-1X as for existing CDMAOne channels, leading to a graceful upgrade. Of course the subscriber units—mobiles—need to be capable of supporting the CDMA2000 specification but this can be done in a more gradual fashion since the existing CDMAOne subscriber units can utilize the new network.

There are several terms used to describe CDMA2000 for the different radio carrier platforms, some of which exist at present and others are in the development phase. However, the sequence of different CDMA2000 platforms or migration path is

CDMA2000-1X

1xEV (1xRTT)

1xDO

1xDV

The 1X uses a single carrier requiring 1.25 MHz of radio spectrum, which is the same as the existing CDMAOne system's channel bandwidth requirement. However, the 1X platforms use a different vocoder and introduction of more Walsh codes, 256/128 versus 64, allowing for higher data rates and more voice conversions than are possible with existing CDMAOne systems.

Under CDMA2000-1X there are three primary methods—1XEV(1xRTT), 1XDO, and 1xDV—which are not mutually exclusive of each other. While the other two are more specific in that 1xDO means one carrier that is data only while 1xDV means one carrier with supports data and voice services, as of this writing, 1xEV is predominantly deployed with some systems introducing 1xDO, 1xDV is still in development.

Another very important aspect of CDMA2000 is that it supports not only IS-41 system connectivity, as does IS-95, but also supports GSM-MAP connectivity requirements leading to the eventual harmonization

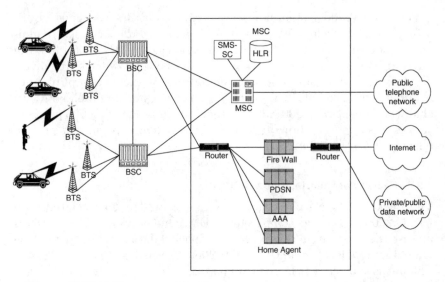

Figure 4.3   CDMA2000 system architecture.

or dual system deployment in the same market by a wireless operator wishing to deploy both WCDMA and CDMA2000 concurrently.

CDMA2000 and IS-95 share many commonalities. However, CDMA2000 requires changes to the radio and network architecture of the existing IS-95 system. A general CDMA2000 network is shown in Fig. 4.3. The connectivity to other like networks is not shown to keep the diagram less cluttered.

The introduction of CDMA2000 into a network requires that major platforms either have upgrades performed or are essentially new to the CDMA2000 network as compared to a CDMAOne system. The platform upgrades involve the BTS and BSC that can be facilitated by module additions or swaps, depending on the infrastructure vendor that is being used. Whether the system is new or upgrading from a CDMAOne system, the heart of the packet data services for a CDMA2000 network is the introduction of the *Packet Data Serving Node* (PDSN).

4.3.1   CDMA radio network

The radio network for IS-95 or CDMA2000 has many similarities. The similarities are required to ensure backward compatibility. The following descriptions regarding the radio network of a CDMA system will use materials directly associated with the CDMA2000 specification. However, it is very important to note that the RC1 and RC2 are associated directly with IS-95 systems.

The CDMA2000 system has several enhancements over existing IS-95/J-STD-008 wireless systems, namely packet data services. In addition to the introduction of packet data services, CDMA2000 has the following general enhancements over IS-95—better power control, diversity transmit, modulation scheme changes, new vocoders, uplink pilot channel, expansion of the existing Walsh codes, and channel bandwith changes. The CDMA2000 radio system following the IS-2000 specification is designed to allow an existing CDMAOne operator a phased entrance into the 3G arena.

4.3.2 Packet data serving node (PSDN)

The Packet Data Serving Node is a new component associated with CDMA2000 systems as compared to CDMAOne networks. The PSDN is an essential element in the treatment of packet data services that will be offered and its location in the CDMA2000 network is shown in Fig. 4.3. The purpose of the PDSN is to support packet data services and it performs the following major functions in the course of a packet data session.

1. Establishes, maintains, and terminates PPP sessions with the subscriber

2. Supports both simple and Mobile IP packet services

3. Establishes, maintains, and terminates the logical links to the Radio Network (RN) across the R-P interface

4. Initiates authentication, authorization, and accounting (AAA) for the mobile station client to the AAA server

5. It receives service parameters for the mobile client from the AAA server

6. Routes packets to and from the external packet data networks

7. Collects usage data that are relayed to the AAA server

The overall capacity of the PSDN is determined by both the throughput and the number of PPP sessions that are being served. The specific capacity of the PSDN is of course dependant upon the infrastructure vendor used as well as the particular card population implemented. It is important to note that capacity is only one aspect of the dimensioning process and that the overall network reliability factor must be addressed in the dimensioning process.

Authentication, authorization, and accounting (AAA). The authentication, authorization, and accounting server is another new component associated with CDMA2000 deployment. The AAA provides, as its names implies, authentication, authorization, and accounting functions for the packet data network associated with CDMA2000 and uses the RADIUS protocol.

The AAA server, as shown in Fig. 4.3, communicates with the PSDN via IP and performs the following major functions in a CDMA2000 network.

- Authentication associated with PPP and Mobile IP connections
- Authorization (service profile and security key distribution and management)
- Accounting

Home agent. The Home Agent (HA) is the third major component to the CDMA2000 Packet Data Service Network and should be compliant with IS-835 that are relevant to the HA functionality within a wireless network. The HA performs many tasks, some of which are tracking the location of the Mobile IP subscriber as it moves from one packet zone to another. The HA in tracking the mobile will also ensure that the packets are forwarded to the mobile itself.

Router. The router shown in Fig. 4.3 has the function of routing packets to and from the various network elements within a CDMA2000 system. The router is also responsible for sending and receiving packets to and from the internal network to the off-net platforms. A firewall, not shown in the figures, is needed to ensure that security is maintained when connecting to off-net data applications.

Home location register (HLR). The HLR used in existing IS-95 networks needs to store additional subscriber information associated with the introduction of packet data services. The HLR performs the same role for packet services as it currently does for voice services in that it stores the subscriber packet data service options and terminal capabilities along with the traditional voice platform needs. The service information from the HLR is downloaded in the *Visitor Location Register* (VLR) of the associated network, switch, during the successful registration process much the same as it is done in existing IS-95 systems and other 1G and 2G voice-oriented systems.

4.3.3 RAN

The Radio Access Network associated with CDMA2000 consists of the subscriber, the *Base Transmitter Site* (BTS), and the *Base Site Controller* (BSC). The subscriber unit can use a variety of services provided the wireless operator makes them available. The inclusion of packet data services has the potential of making the subscriber experience more ubiquitous.

BSC. The BSC is responsible for controlling all of the BTSs under its domain. The BSC routes packets to and from the BTSs to the PDSN. In addition, the BSC routes TDM traffic to the circuit-switched platforms and packet data to the PDSN.

4.3.4 BTS

The BTS is the official name of the cell site. The BTS is responsible for allocating resources, both power and Walsh codes, for consumption by the subscribers. The BTS also has the physical radio equipment that is used for transmitting and receiving the CDMA2000 signals.

The BTS controls the interface between the CDMA2000 network and the subscriber unit. The BTS controls many aspects of the system that are directly related to the performance of the network. Some of the items the BTS controls are the multiple carriers that operate from the site, the forward power (allocated for traffic, overhead, and soft handoffs), and of course the assignment of the Walsh codes.

Therefore, when a new voice or packet session is initiated the BTS must decide how to best assign the subscriber unit to meet the services being delivered. The BTS in the decision process not only examines the service requested, but also must consider the radio configuration and the subscriber type and of course whether the service requested is voice or packet. Therefore, the resources the BTS has to draw upon can be both physically and logically limited depending on the particular situation involved.

The physical resources the BTS draws upon also involve the management of the channel elements that are required for both voice and packet data services. In addition, handoffs are accepted or rejected on the basis of the available power only.

Integral to the resource assignment scheme is the Walsh code management. For 1X, whether 1xEV, 1xDO, or 1xDV, there are a total of 128 Walsh codes to draw upon. However, the Walsh codes in later releases can be expanded to a total of 256. The inclusion of the legacy system IS-95 A/B or J-STD-008 factors in the Walsh codes as well, in that these systems can only support 64 codes.

For CDMA2000-1x the voice and data distribution is handled by parameters that are set by the operator, which involve

Data resources (percent of available resources which includes FCH and SCH)

FCH resources (percent of data resources)

Voice resources (percent of total available resources)

This is an important concept, as with all RAN resources, in that the operator can and does determine the resources that will be allocated for both circuit and packet services, i.e., voice and data. In other words, the allocation of data/FCH resources directly controls the number of simultaneous data users on a particular sector or cell site.

SR and RC.   CDMA2000 defines two *spreading rates* (SR) as compared to IS-95 that has only one spreading rate. The spreading rates are referred to as SR1 for spreading rate 1 and SR3 for spreading rate 3. The SR1 spreading rate is used for IS-95A/B and CDMA2000 phase 1, 1xRTT implementations while SR3 is destined for CDMA2000 Phase 2, 3xRTT.

The IS-2000 specification defines 1xRTT radio access methods to include a total of nine forward and six reverse links radio configurations (RC) as well as two different spreading rates. The RC involve different modulations, coding, and vocoders, while the spreading rates address the amount of use of the two different chip rates. The radio configurations are referred to as RC1 for radio configuration 1.

RC1 is backward compatible with CMDAOne for 9.6 kbps voice traffic and supports circuit-switched data rates of 1.2 kbps to 9.6 kbps. While RC3 is based on the 9.6 kbps rate and supports variable voice rates from 1.2 k to 9.6 kbps, while also supporting packet data rates of 19.2, 38.4, 76.8, and 153.6 kbps, it operates using an SR1.

The following two tables are meant to help illustrate the perturbations that exist with the different radio configurations and spreading rates. Table 4.2 is associated with the forward link while Table 4.3 is associated with the reverse link.

Walsh codes.   CDMA2000 introduces an increase in the number of Walsh codes from 64 (with IS-95) to 128 and eventually to 256. As with IS-95, CDMA2000 uses PN long codes for both the forward and reverse direction. However, in CDMA2000 the introduction of variable length Walsh

TABLE 4.2   Forward Link RC and SR

RC	SR	Data rates	Characteristics
1	1	1200, 2400, 4800, 9600	$R = 1/2$
2	1	1800, 3600, 7200, 14400	$R = 1/2$
3	1	1500, 2700, 4800, 9600, 38400, 76800, 153600	$R = 1/4$
4	1	1500, 2700, 4800, 9600, 38400, 76800, 153600, 307200	$R = 1/2$
5	1	1800, 3600, 7200, 14400, 28800, 57600, 115200, 230400	$R = 1/4$
6	3	1500, 2700, 4800, 9600, 38400, 76800, 153600, 307200	$R = 1/6$
7	3	1500, 2700, 4800, 9600, 38400, 76800, 153600, 307200, 614400	$R = 1/3$
8	3	1800, 3600, 7200, 14400, 28800, 57600, 115200, 230400, 460800	$R = 1/4$ (20 ms) $R = 1/3$ (5 ms)
9	3	1800, 3600, 7200, 14400, 28800, 57600, 115200, 230400, 460800,1036800	$R = 1/2$ (20 ms) $R = 1/3$ (5 ms)

TABLE 4.3 Reverse Link RC and SR

RC	SR	Data rates	Characteristics
1	1	1200, 2400, 4800, 9600	$R = 1/3$
2	1	1800, 3600, 7200, 14400	$R = 1/2$
3*	1	1200, 1350, 1500, 2400, 2700, 4800, 9600, 19200, 38400, 76800, 153600, 307200	$R = 1/4$ $R = 1/2$ for 307200
4*	1	1800, 3600, 7200, 14400, 28800, 57600, 115200, 230400	$R = 1/4$
5*	3	1200, 1350, 1500, 2400, 2700, 4800, 9600, 19200, 38400, 76800, 153600, 307200, 614400	$R = 1/4$ $R = 1/2$ for 307200 and 614400
6*	3	1800, 3600, 7200, 14400, 28800, 57600, 115200, 230400, 460800, 1036800	$R = 1/4$ $R = 1/2$ for 1036800

codes is made to accommodate fast packet data rates. The Walsh code chosen by the system is determined by the type of reverse channel.

The inclusion of packet data services for CDMA2000 required the increased Walsh codes. What is important about the Walsh codes is not only the data rate associated with the various codes but also the interrelationship the codes have with the data throughput.

Table 4.4 shows the relationship between Walsh codes, SR, RC, and, of course, data rates. An important issue, or rather effect of using variable length Walsh codes, is that if a shorter Walsh code is being used then it precludes the use of the longer Walsh codes that are derived from it.

Table 4.4 helps in establishing the relationship between Walsh code lengths and associated data rates. It is important to note that the shorter Walsh codes inhibit the use of longer Walsh codes because of the orthogonality required. Also, all channel requests are allocated from the same Walsh code pool on a per sector basis. In addition, to achieve the higher

TABLE 4.4 Walsh Code Tree Table

		Walsh codes						
	RC	256	128	64	32	16	8	4
SR1	1	NA	NA	9.6	NA	NA	NA	NA
	2	NA	NA	14.4				
	3	NA		9.6	19.2	38.4	76.8	153.6
	4	NA	9.6	19.2	38.4	76.8	153.6	307.2
	5	NA	NA	14.4	28.8	57.6	115.2	230.4
SR3	6		9.6	19.2	38.4	76.8	153.6	307.2
	7	9.6	19.2	38.4	76.8	153.6	307.2	614.4
	8		14.4	28.8	57.6	115.2	230.4	460.8
	9	14.4	28.8	57.6	115.2	230.4	460.8	1036.8

data rate not only is the Walsh codes implementation modified but also the modulation scheme has been changed.

4.3.5 Call and data processing

For CDMA2000 there are several types of call and data processing that take place. Depending on whether the service being requested or offered is circuit-switched or packet, the call processing methods are different. Regardless of whether the service being delivered is circuit- or packet-based the system still will perform handoffs in addition to power control. Granted, there are a few nuances to be concerned about depending on whether the service is packet or voice, but the fundamental concepts of how neighbors are promoted and demoted is still the same.

What follows over the next few sections are discussions, with flowcharts, regarding various circuit-switched and packet calls.

Call processing. The call flows for CDMA are shown in Figs. 4.4 and 4.5. It is important to note that IS-95 is primarily a voice and not a data

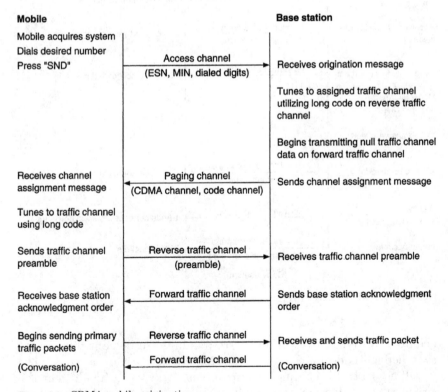

Figure 4.4 CDMA mobile origination.

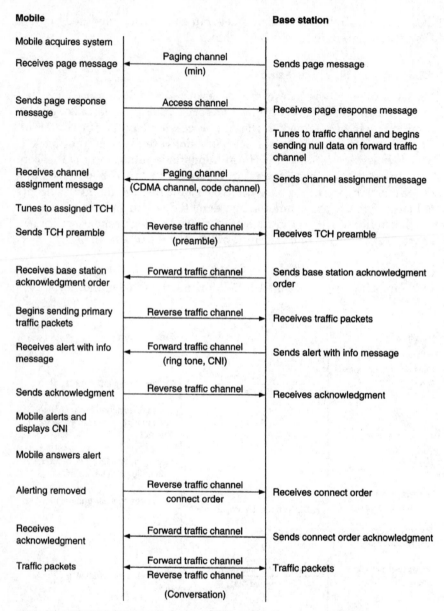

Figure 4.5   CDMA mobile termination.

oriented system. However, data are available to be sent via circuit-switched methods but the call processing flow is the same as voice since it still uses a traffic channel, set up for voice transport. In addition, the call processing flowcharts that follow can easily be expanded to include the variations of subscriber types for Walsh code allocations along with different spreading rates.

The first call processing flowchart, Fig. 4.4, is for a mobile to land call (origination), while the second flowchart, Fig. 4.5, illustrates a land to mobile call (termination).

Packet data transport process flow. CDMA2000 data services fall within two distinct categories, circuit switched and packet. Circuit-switched data are handled effectively the same as a voice call. But for all packet data calls a PDSN is used as the interface between the air interface data transport and the fixed network transport. The PDSN interfaces to the base station (BS) through a *Packet Control Function* (PCF) that can be colocated with the BS.

The CDMA2000 has three packet data service states that need to be understood in the process–: active/connected, dormant, and null/inactive.

Active/connective packet data. A physical traffic channel exists between the subscriber unit and the base station with packet data being sent and received in a bidirectional fashion.

Dormant. No physical traffic channel exists but a PPP link between the subscriber unit and the PDSN is maintained.

Null/inactive. The third possible situation, in this case neither a traffic channel nor PPP link is maintained or established.

The relationship between the three packet data states is best shown in the simplified state diagram in Fig. 4.6.

CDMA2000 introduces to the mobility environment real packet data transport and treatment at speeds that meet or exceed the ITM-2000

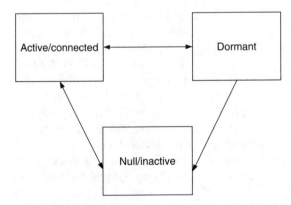

Figure 4.6 Packet data states.

system requirements. The voice call processing that is implemented by CDMA2000 is functionally the same as that of existing CDMAOne, IS-95, networks with the exception that there is a vocoder change in the subscriber units. However, the key difference is that packet data can now be handled by the network.

The mobile initiates the decision as to whether the session will be a packet data session, voice session, or concurrent, meaning voice and data. The network at this time cannot initiate a packet data session with the subscriber unit with the exception of SMS.

For call processing the voice and data networks are segregated in general once the information, whether it is voice or data, leaves the radio environment at the BSC itself. Therefore, for packet data the PDSN is central to all decisions that are made. Refer to Fig. 4.3, which depicts a generalized network architecture.

The PDSN does not communicate directly with the voice network nodes like the HLR and VLR instead it is done via the AAA. As discussed earlier the voice and data networks normally are segregated once they leave the radio environment at the BSC. Additionally, in a CDMA2000 network the system uses a Point-to-Point (PPP) Protocol between the mobile and the PDSN for every type of packet data session that is transported and or treated.

The PDSN is meant to provide several key packet data services and they are Simple IP and Mobile IP. There are also several variants, to be discussed shortly, relative to each of these services. However, the concepts of Simple IP and Mobile IP need to be explored first.

Simple IP is a packet data service relative to CDMA2000, 1xRTT, and is where the subscriber is assigned a *Dynamic Host Configuration Protocol* (DHCP) address from the serving PSDN with its routing service provided by the local network. The specific IP address, which the subscriber is assigned, remains with the subscriber when being served by the same radio network that maintains connectivity with the PSDN that issued the IP address. It is important to note that Simple IP does not provide for mobile terminations and therefore is an origination-based service only, that is, a PPP service using DHCP.

Mobile IP is the other packet data service relative to CDMA2000. Mobile IP is where the mobile's IP routing service is provided by the public IP network and for this functionality the mobile is assigned a static IP address that resides with the Home Agent (HA). A key advantage of Mobile IP over Simple IP is that the mobile, due to the static IP address, can handoff between different radio networks that are served via different PSDNs, which resolves the roaming issues that are part of Simple IP. Mobile IP also enables the possibility for mobile terminations due to the static IP.

Now, with mobility whether the packet service is Simple IP or Mobile IP, the notion of mobility is fundamental to the concept of CDMA2000.

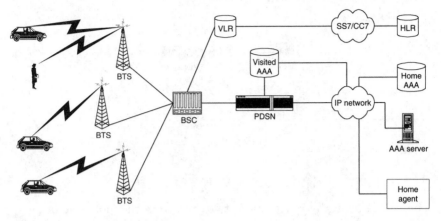

Figure 4.7   CDMA2000 Packet Network nonhome.

Therefore, the diagram shown in Fig. 4.7 illustrates some of the inter-network communication that needs to take place in establishing a packet data session.

It is important to note that the transport of the packets is not depicted in Fig. 4.7; just the elements in the network that need to communicate in order to establish what services the subscriber is allowed to have and how the network is going to meet the SLA that is expected for the packet session.

The VLR, which is normally colocated with the MSC, is shown in Fig. 4.7. When the subscriber initiates a packet data session the BSC via the MSC checks the subscriber subscription information prior to the system granting the service request to the mobile subscriber. This will take place prior to the PDSN being involved with the packet session.

Elaborating on the various packet sessions available for use within a CDMA2000 network are, of course, Simple and Mobile IP. However, under each of these packet session types are two variants: One uses a *Virtual Private Network* (VPN) And the other does not.

- Simple IP
- Simple IP-VPN
- Mobile IP
- Mobile IP-VPN

A more specific discussion of Simple IP and Mobile IP is covered next.

Simple IP.   Simple IP is very similar to the dial-up Internet connections used by many people over standard landline facilities. Simple IP is where a PPP session is established between the mobile and the PDSN.

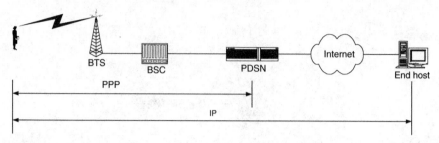

Figure 4.8 Simple IP.

The PDSN basically routes packets to and from the mobile in order to provide end-to-end connectivity between the mobile and the Internet. A diagram depicting Simple IP is shown in Fig. 4.8.

When using Simple IP the mobile must be connected to the same PSDN for the duration of the packet session. If the mobile while in transit moves to a coverage area whose BSC/BTSs are homed out of another PSDN the Simple IP connection is lost and needs to be reestablished. The loss of the existing packet session effectively is the same as when the Internet connection on the landline is terminated and you need to reestablish the connection.

Refer to Fig. 4.7, which is a simplified model of the Simple IP implementation. Many of the details are left out but the concept shows that the mobile is connected to the PDSN using a PPP connection in a best-effort data delivery method at whatever the agreed upon transfer rate that was negotiated, which is determined by the subscriber's profile, radio resource availability, and of course the radio environment itself.

The IP address of a mobile is linked to the PDSN that can be static or DHCP (for Simple IP the choice is DHCP). A mobile with an active or dormant data call can transverse around the network going from cell to cell provided it stays within the PDSN's coverage area. Additionally, the PSDN should support both *Challenge Handshake Authentication Protocol* (CHAP) and *Password Authentication Protocol* (PAP).

Simple IP, as indicated, does not allow the subscriber full mobility with packet data calls. When the subscriber exits the PDSN coverage area, it must negotiate for a new IP address from the new PDSN, which of course results in the termination of the existing packet session and requires a new session to begin.

Regarding the radio environment, the CDMA2000 radio network provides the mobile with a traffic channel that consists of a fundamental channel and possibly a supplemental channel for higher traffic speeds. To help explain the use of the Simple IP process, a call flow diagram

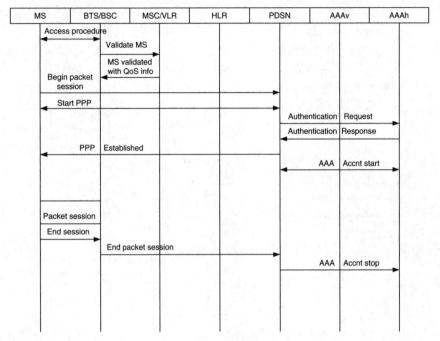

MS	BTS/BSC	MSC/VLR	HLR	PDSN	AAAv	AAAh

Figure 4.9   Simple IP flowchart.

(packet session flowchart) is shown in Fig. 4.9 that represents the situation when a subscriber is operating in a home PDSN network. In Fig. 4.10, the mobile is considered to be roaming.

Simple IP with VPN.   An enhancement to Simple IP is the ability to introduce a VPN to the path for security and also to provide connectivity to corporate LAN or other packet networks.

With VPN the mobile user should appear to be connected directly to the corporate LAN.

The PDSN establishes a tunnel using L2TP Protocol between the PDSN and the Private Data Network. The mobile is effectively still using a PPP connection but it is tunneled. The private network that the PDSN terminates to is responsible for assigning the IP address and of course authenticating the user beyond what the wireless system needs to perform for billing purposes.

Because of the specific termination and authentication that is performed by another network, the PDSN does not apply any IP services for the mobile and, therefore, except for the predetermined speed of the connection, that is all the system can provide.

Just as in the case of Simple IP, the mobile must still be connected to the same PDSN for the packet session. If the mobile moves to another

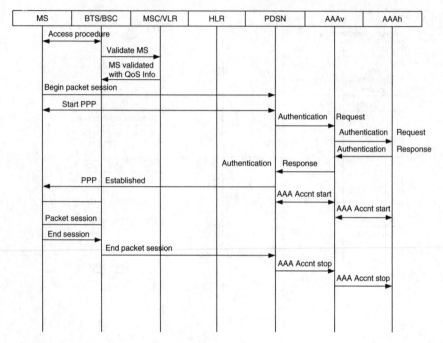

Figure 4.10 Simple IP roaming flowchart.

area of the network, which is covered by a separate PDSN, the VPN is terminated and the mobile must reestablish the session.

A simplified diagram is shown in Fig. 4.11.

The packet session flowchart for Simple IP VPNM is shown in Fig. 4.12 and assumes that the subscriber is *not* roaming.

Mobile IP (3G).   Mobile IP, although a packet transport method, is quite different from Simple IP in that it actually transports data. Mobile IP uses a static IP address that can be assigned by the PDSN. The establishment

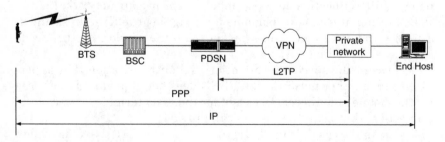

Figure 4.11 Simple IP–VPN.

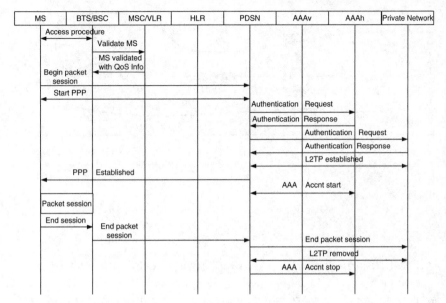

Figure 4.12   Simple IP with VPN session flowchart.

of a static IP address facilitates roaming during the packet session, provided the static IP address scheme is unique enough for the subscriber unit to be uniquely identified.

With Mobile IP the PDSN is the Foreign Agent (FA) and the Home Agent is set up as a virtual HA. The mobile needs to register each time it begins a packet data session, whether it is originating or terminating. Also, the PDSN on the visited network terminates the packet session using an IP in IP tunnel. The HA delivers the IP traffic to the FA through an IP tunnel.

The mobile is responsible for notifying the system that it has moved to another service area. Once the mobile has moved to another service area it needs to register with another FA. The FA assigns the mobile a care of address (COA). The HA forwards the packets to the visited network for termination on the mobile. The HA encapsulates the original IP packet destined for the mobile using the COA. The FA using IP in IP tunneling extracts the original packet and routes it to the mobile.

The IP address assignment is done via DHCP and is mapped to the HA. However, PAP and CHAP are not used for Mobile IP as in Simple IP.

In the reverse direction, the routing of IP packets occurs the same as if on the home network and does not require an IP in IP tunnel unless the wireless operator decides to implement reverse IP tunneling.

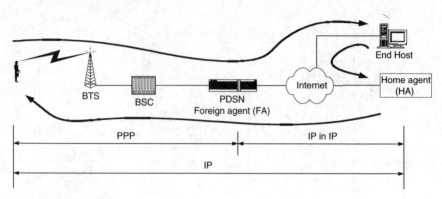

Figure 4.13 Mobile IP.

In summary,

- The PDSN in the visited network always terminates the IP in IP tunnel.

- The HA delivers the IP traffic through the Mobile IP tunnel to the FA.

- The FA performs the routing to the mobile and assigns the IP address using DHCP.

The diagram shown in Fig. 4.13 is a simplified depiction of Mobile IP. What follows next is an example of a Mobile IP packet session flow that is shown in Fig. 4.14.

Mobile IP with VPN. The second variant to Mobile IP is Mobile IP with VPN. Mobile IP with VPN affords greater mobility for the subscriber over Simple IP with VPN since it can maintain a session when it moves from one PDSN area to another. Like Mobile IP, the IP address assigned to the subscriber is static. However, the private network the mobile is connected to provides the IP address that needs to be drawn from a predefined IP scheme that is coordinated. The PDSN provides a COA when operating in a nonhome PDSN for routing purposes. The IP packets in both directions, however, flow between the HA and the FA using IP in IP encapsulation and no treatment, with the exception of throughput speed allowed, is performed by the wireless network.

The brief diagram shown in Fig. 4.15 depicts the general packet flow for Mobile IP with VPN.

4.3.6 Handoffs

There are several types of handoffs available with CDMA. The types of handoffs involve soft, softer, and hard. The difference between the types is dependent on what is sought to be accomplished.

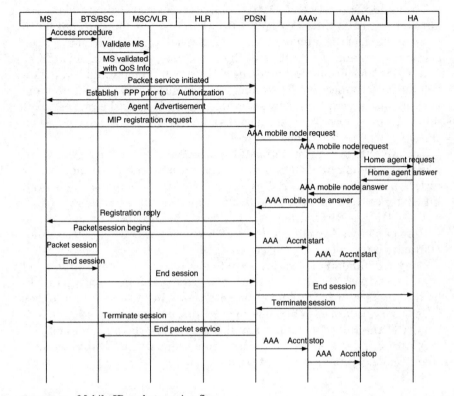

Figure 4.14 Mobile IP packet session flow.

There are several user adjustable parameters that help the handoff process take place. The parameters that need to be determined involve the values to add or remove a pilot channel from the active list and the search window sizes. There are several values that determine when to add or remove a pilot from consideration. In addition, the size of the search window cannot be too small nor can it be too large.

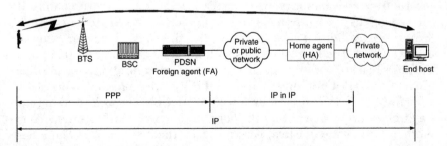

Figure 4.15 Mobile IP with VPN.

The handoff process for CDMA can take on several variants. The variants for handoffs in CDMA are soft handoff, softer handoff, and hard handoff. Each of the handoff scenarios is a result of the particular system configuration and the location of the subscriber unit in the network.

In CDMA a soft handoff involves an intercell handoff and is a make-before-break connection. The connection between the subscriber unit and the cell site is maintained by several cell sites during the process. Soft handoff can occur only when the old and new cell sites are operating on the same CDMA frequency channel.

The advantage of a soft handoff is path diversity for the forward and reverse traffic channels. Diversity on the reverse traffic channel results in less power being required by the mobile unit, reducing the overall interference that increases the traffic handling capacity.

The CDMA softer handoff is an intracell handoff occurring between sectors of a cell site and is a make-before-break type. The softer handoff occurs only at the serving cell site.

The hard handoff process is meant to enable a subscriber unit to hand from a CDMA call to an analog call. The process is functionally a break-before-make and is implemented in areas where there is no longer CDMA service for the subscriber to use while on a current call. The continuity of the radio link is not maintained during the hard handoff.

A hard handoff can also occur between two distinct CDMA channels that are operating on different frequencies.

4.4 GSM

Global System for Mobile Communications (GSM) has many unique features and attributes that make it an excellent digital radio standard to use, resulting in it being the most widely accepted radio communication standard at this time. GSM was developed in Europe as a communication standard that would be used throughout Europe in response to the problem of multiple and incompatible standards that still exist there today.

The GSM radio channel is 200 kHz wide. GSM has been deployed in several frequency bands namely the 900, 1800, and 1900 MHz bands. It consists of the following major building blocks, the *Switching System* (SS), the *Base Station System* (BSS), and the *Operations and Support System* (OSS). The BSS comprises both the Base Station Controller and the Base Transceiver Stations. In an ordinary configuration several BTSs are connected to a BSC and then several BSCs are connected to the MSC.

Figure 4.16 shows the basic architecture of a GSM network. Working our way from the left, we see that the handset, known in GSM as the *Mobile Station* (MS), communicates over the air interface with a BTS. The MS is composed of two parts: the handset itself, known as the *Mobile Equipment*

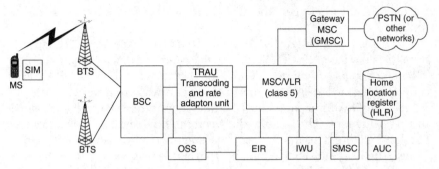

Figure 4.16   Generic GSM system.

(ME), and the *Subscriber Identity Module* (SIM), a small card containing
an integrated circuit. The SIM contains user-specific information, including the identity of the subscriber, subscriber authentication information,
plus some subscriber service information. It is only when a given subscriber's SIM is inserted into a handset that the handset acts in accor-
dance with the services the subscriber has subscribed to. In other words,
my handset acts as my handset only when my SIM is inserted.

The BTS contains the radio transceivers that provide the radio interface with mobile stations. One or more BTSs are connected to a *Base
Station Controller*. The BSC provides a number of functions related to
radio resource (RR) management, some functions related to *mobility
management* (MM) for subscribers in the coverage area of the BTSs, and
a number of operation and maintenance functions for the overall radio
network. Together, BTSs and BSCs are known as the *Base Station
Subsystem* (BSS).

The interface between the BTS and the BSC is known as the *Abis interface*. Many aspects of that interface are standardized. One aspect, however, is proprietary to the BTS and BSC vendor. That is the part of the
interface that deals with configuration, operation, and maintenance of the
BTSs. This is known as the *Operation and Maintenance Link* (OML).
Since the internal design of a BTS is proprietary to the BTS vendor, and
since the OML needs to have functions that are specific to that internal
design, the OML is also proprietary to the BTS vendor. The result is
that a given BTS must be connected to a BSC of the same vendor.

One or more BSCs are connected to an MSC. The MSC is the "switch":
the node that controls call setup, call routing, and many of the functions
provided by a standard telecommunications switch. The MSC needs to provide a number of MM functions. It also needs to provide a number of interfaces that are unique to the GSM architecture.

The VLR is a database that contains subscriber-related information
for the duration that a subscriber is in the coverage area of an MSC. The
MSC and VLR are always contained on the same platform.

The interface between the BSC and the MSC is known as the *A-interface*.

The *Transcoding and Rate Adaptation Unit* (TRAU) deals with the speech from the subscriber. The TRAU is usually coded at either 13 kbps (*full rate*, FR) or 12.2 kbps (*enhanced full rate*, EFR). The function of the TRAU is to convert the coded speech to/from standard 64 kbps to interface with the PSTN. The TRAU is a part of the BSS functionality and is usually contained in a separate rack. The interface between the BSC and TRAU is known as the *Ater interface*. This interface is proprietary to the BSS equipment vendor. Hence, the BSC and TRAU must be from the same vendor.

In Fig. 4.16 we also find an HLR—a node found in most, if not all, mobile networks. The HLR contains subscriber data, such as the details of the services to which a user has subscribed. Associated with the HLR, we find the *Authentication Center* (AuC). This is a network element that contains subscriber-specific authentication data, such as a secret authentication key called the *Ki*. The AuC also contains one or more sophisticated authentication algorithms. For a given subscriber, the algorithms in the AuC and the Ki are also found on the SIM card. Using a random number assigned by the AuC and passed down to the SIM via the HLR, MSC, and ME, the SIM performs a calculation using the Ki and authentication algorithm. If the result of the calculation on the SIM matches that in the AuC, the subscriber has been authenticated. The interface between the HLR and AuC is not standardized. Although there are implementations where the HLR and AuC are separate, it is more common to find the HLR and AuC integrated on the same platform.

Calls from another network, such as the PSTN, first arrive at a type of MSC known as a *Gateway MSC* (GMSC). The main purpose of the GMSC is to query the HLR to determine the location of the subscriber. The response from the HLR indicates the MSC where the subscriber may be found. The call is then forwarded from the GMSC to the MSC serving the subscriber. A GMSC may be a full MSC/VLR such that it may have some BSCs connected to it. Alternatively, it may be a dedicated GMSC whose only function is to interface with the PSTN. The choice is dependent on the amount and types of traffic in the network and the relative cost of a full MSC/VLR versus a pure GMSC.

The *Short Message Service Center* (SMSC) in Fig. 4.16 is a node that supports the store-and-forward of short messages to/from mobile stations. Typically, though not always, these short messages are text messages up to 160 characters in length.

The *Equipment Identity Register* (EIR) is shown in Fig. 4.16. The EIR verifies that a particular handset (ME) or model of ME the subscriber wants to use is acceptable on the system. Therefore, to some degree, the handset used by a particular subscriber is not relevant. Stored in each handset is an *International Mobile Identity* number (IMEI, 15 digits) or the *International Mobile Equipment Identity and Software Version Number* (IMEISV16 digits). Both the IMEI and IMEISV have a structure that includes the *type approval code* (TAC) and the *final assembly code* (FAC). The TAC and FAC combine to indicate the make/model of the handset and the place of manufacture. The IMEI and IMEISV also include a specific serial number for the ME in question. The only difference between IMEI and IMEISV is the software version number. Within the EIR, there are three lists: black, gray, and white. These lists contain values of TAC, TAC and FAC, or complete IMEI or IMEISV. If a given TAC, TAC/FAC combination, or complete IMEI appears in the black list, then calls from the ME are barred. If it appears in the gray list, then calls may or may not be barred at the discretion of the network operator. If it appears in the white list, then calls are allowed. Typically, a given TAC is included in the white list if the model of handset is one that has been approved by the handset manufacturer. The EIR is an optional network element and some network operators have chosen not to deploy an EIR.

The *Interworking Function* (IWF) is used for circuit-switched data and fax services and is basically a modem bank. Typical dial-up modems and fax machines are analog. Therefore, a circuit-switched data call from an MS is looped through the IWF before being routed onward by the IWF. Within the IWF, a modem is placed in the call path. The same applies for facsimile service, where a fax modem would be used rather than a data modem. GSM supports data and fax services up to 9.6 kbps.

4.4.1 GSM RAN

The GSM is a *Time Division Multiple Access* (TDMA) system that uses *Gaussian Minimum Shift Keying* (GMSK) as the modulation scheme. A GSM RAN consists of several 200 kHz carriers or RF channels in both the uplink and downlink.

In GSM a given band is divided into a number of RF channels or carriers, each 200 kHz in both the uplink and downlink. Thus, if a handset is transmitting on a given 200 kHz carrier in the uplink, then it is receiving on a corresponding 200 kHz carrier in the downlink. Since the uplink and downlink are rigidly associated, when one talks about a carrier or RF channel both the uplink and downlink are usually implied. A given cell can have multiple RF carriers—typically one to three in a normally loaded system, though as many as six carriers might exist in a heavily loaded cell in an area of very high traffic demand. Typically, in GSM a cell is referred to as a sector. Thus, a three-sector BTS implies three cells.

4.4.2 Location update

When an MS is first turned on, it must first "camp on" a suitable cell. This largely involves scanning the air interface to select a cell with a strong

received signal strength and decoding the information broadcast by the
BTS on the BCCH. Generally, the MS will camp on the cell with the
strongest signal strength, provided that the cell belongs to the *Home
Public Land Mobile Network* (HPLMN) and provided that the cell is not
barred. The MS then registers with the network, which involves a process
known as *location updating* as shown in Fig. 4.17.

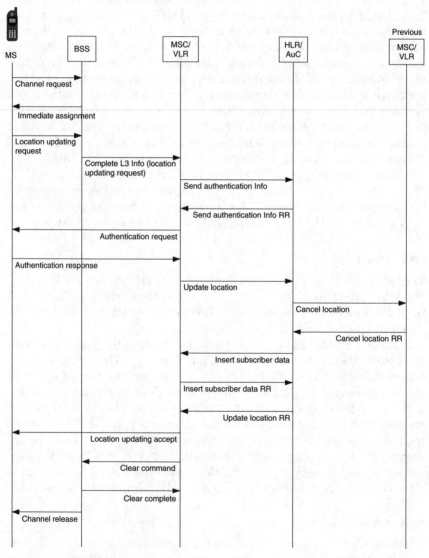

Figure 4.17 GSM location update.

4.4.3 Mobile-originated voice call

Figure 4.18 shows a basic mobile-originated call to the PSTN. After the MS has been placed on a *Stand-alone Dedicated Control Channel* (SDCCH) by the BSS (not shown), the MS issues a *Connection Management (CM) Service Request* to the MSC. This includes information about the type of service that the MS wishes to invoke (a mobile-originated call in this case, but could also be another service such as SMS). On receipt of the CM Service Request, the MSC may optionally invoke authentication of the mobile. The MSC initiates ciphering so that the voice and data sent over the air are encrypted. Since it is the BSS that performs the encryption and decryption, the MSC needs to pass the Kc to the BSS. The BSS then instructs the MS to start ciphering. The MS, of course, generates the Kc independently, so that it is not passed over the air. Once the MS has started ciphering, it informs the BSS, which in turn informs the MSC.

4.4.4 Mobile-terminated voice call

A basic mobile-terminated call from the PSTN is shown in Fig. 4.19.

4.4.5 Handover

Handover (also known as handoff) is the process by which a call in progress is transferred from a radio channel in one cell to another radio channel, either in the same cell or in a different cell. Handover can occur within a cell, between cells of the same BTS, between cells of different BTSs connected to the same BSC, between cells of different BSCs, or between cells of different MSCs. Not only can handover occur between TCHs, it is also possible from an SDCCH on one cell to an SDCCH on another cell, and is possible from an SDCCH on one cell to a TCH on another cell. The most common, however, is handover from TCH to TCH.

Depending on the source (i.e., original cell) and target (i.e., destination cell) involved in the handover, the handover may be handled completely within a BSS or may require the involvement of an MSC. In the case where a handover occurs between cells of the same BSC, the BSC may execute the handover and simply inform the MSC after the handover has taken place. If, however, the handover occurs between BSCs then the MSC must become involved, since there is no direct interface between BSCs.

Handover in GSM is known as *Mobile-Assisted Handover* (MAHO). This means that it is the network that decides if, when, and how a handover should take place. The MS, however, provides information to the network to enable the network to make the decision.

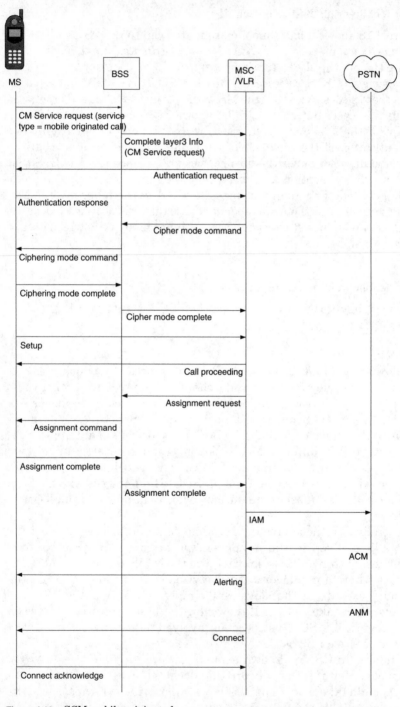

Figure 4.18 GSM mobile originated.

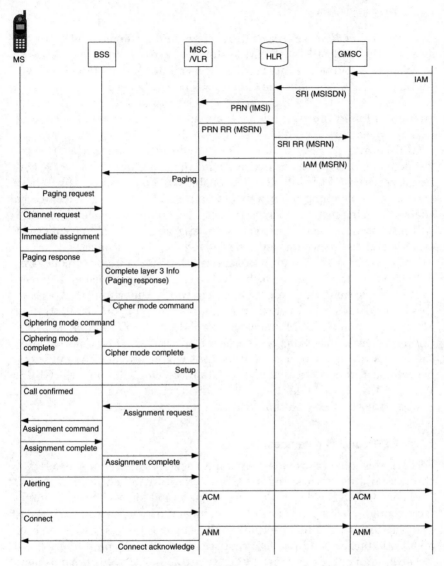

Figure 4.19 GSM mobile terminated.

In addition to the measurements reported by the MS, the BTS itself makes measurements regarding the *Receive Level* (RXLEV) and *Receive Quality* (RXQUAL) received from the MS. These measurements and those from the MS are reported to the BSC. Based on its internal algorithms, the BSC makes the decision as to whether a handover should occur, and if so, to which cell.

4.5 GPRS

General Packet Radio Service (GPRS) is an enhancement to GSM type systems that will allow a GSM system to access a subscriber's LAN, WAN, and of course the Internet. GPRS is an important step in the evolution to 3G mobile telephone networks. The key benefit of GPRS is that it integrates higher throughput packet data to mobile networks, fully enabling Mobile Internet applications and a range of other advanced data services.

GPRS provides GSM operators the ability to offer customers better wireless access to the Internet as well as a wide range of other IP-based, packet-based, services. GPRS will be able to enable GSM operators to offer a host of IP services including email, web browsing, and, of course, an e-commerce enabler from the customer perspective. GSM operators implementing GRPS will need to implement various system upgrades and enhancements in order to enable this technology to be delivered to the customers.

GSM provides voice and data services that are circuit-switched. For data services, the GSM network effectively emulates a modem between the user device and the destination data network. Unfortunately, however, this is not necessarily an efficient mechanism for support of data traffic. Moreover, standard GSM supports user data rates of up to 9.6 kbps. In this age of the Internet such a speed is considered very slow. Consequently, the need was obvious for a solution that would provide more efficient packet-based data services at higher data rates. One solution is the GPRS. While GPRS does not offer the high bandwidth services envisioned for 3G, it is an important step in that direction.

4.5.1 GPRS packet data rates

GPRS is designed to provide packet data services at higher speeds than those available with standard GSM circuit-switched data services. In theory, GPRS could provide speeds of up to 171 kbps over the air interface, though such speeds are never achieved in real networks. In fact, the practical maximum is actually a little over 100 kbps, with speeds of about 40 kbps or 53 kbps more realistic with good RF planning.

The greater speeds provided by GPRS are achieved over the same basic air interface as GSM. With GPRS, however, the MS can have access to more than one timeslot. Moreover, the channel coding for GPRS is somewhat different to that of GSM. In fact, GPRS defines a number of different channel coding schemes. The most commonly used coding scheme for packet data transfer is *Coding Scheme 2* (CS-2), which allows for a given timeslot to carry data at a rate of 13.4 kbps. If a single user has access to multiple timeslots then speeds such as 40.2 kbps or 53.6 kbps become available to that user. Table 4.5 lists the various coding schemes available and the associated data rates for a single timeslot.

TABLE 4.5   GPRS Coding Schemes and Data Rates

Coding scheme	Air interface data rate (kbps)	Approximate usable data rate (kbps)
CS-1	9.05	6.8
CS-2	13.4	10.0
CS-3	15.6	11.7
CS-4	21.4	16.0

For a given amount of data to be transmitted, smaller application packet sizes cause a greater net overhead than larger packet sizes. The result is that the rate for usable data is approximately 20 to 25 percent less than the air interface rate.

The most commonly used coding scheme for user data is CS-2. This scheme provides reasonably robust error correction over the air interface. While CS-3 and CS-4 provide higher throughput, they are more susceptible to errors on the air interface. In fact, CS-4 provides no error correction at all on the air interface. Consequently, CS-3 and particularly CS-4 will generate a great deal more retransmission over the air interface. With such retransmission, the net throughput may well be no better than that of CS-2.

Of course, the biggest advantage of GPRS is not simply the fact that it allows higher speeds but that it is a packet-switching technology. This means that a given user consumes RF resources only when sending or receiving data. If a user is not sending data at a given instant, then the timeslots on the air interface can be used by another user. Consider, for example, a user who is browsing the web. Data are transferred only when a new page is being requested or sent. Nothing is being transferred while the subscriber contemplates the content of a page. During this time, some other user can have access to the air interface resources, with no adverse impact to our web-browsing friend. Clearly, this is a very efficient use of scarce RF resources.

The fact that GPRS allows multiple users to share air interface resources is a big advantage. This means, however, that whenever a user wishes to transfer data, the MS must request access to those resources and the network must allocate the resources before the transfer can take place. While this appears to be the antithesis of an "always-connected" service, the functionality of GPRS is such that this request-allocation procedure is well hidden from the user and the service appears to be "always-on." Imagine, for example, a user who downloads a web page and then waits for some time before downloading another page. In order to download the new page, the MS requests resources, is granted resources by the network, then sends the web page request to the network, which forwards the request to the external data network (e.g., the Internet). This happens quite quickly and so the delay is not great. Quite soon, the new page appears on the user's device and at no point did the user have to *dial-up* to the ISP.

4.5.2 GPRS devices

GPRS is effectively a packet-switching data service overlaid on the GSM infrastructure, which is primarily designed for voice. Furthermore, while there is certainly a demand for data services, voice is still the big revenue generator, at least for now. Therefore, it is reasonable to assume that users will require both voice and data services and that operators will want to offer such services either separately or in combination. Consequently, there are three classes of GPRS users.

- *Class A.* It supports simultaneous use of voice and data services. Thus, a Class A user can hold a voice conversation and transfer GPRS data at the same time.

- *Class B.* It supports simultaneous GPRS attach and GSM attach, but not simultaneous use of both services. A Class B user can be "registered" on GSM and GPRS at the same time, but cannot hold a voice conversation and transfer data simultaneously. If a Class B user has an active GPRS data session and wishes to establish a voice call, then the data session is not cleared down. Rather, it is placed on hold until such time as the voice call is finished.

- *Class C.* It can attach to either GSM or GPRS, but cannot attach to both simultaneously. Thus, at a given instant, a Class C device is either a GSM device or a GPRS device. If attached to one service, the device is considered detached from the other.

4.5.3 GPRS air interface

The GPRS air interface is built upon the same foundations as the GSM air interface—the same 200 kHz RF carrier and the same eight timeslots per carrier. This allows GSM and GPRS to share the same RF resources. In fact, if one considers a given RF carrier, then at a given instant some of the timeslots may be carrying GSM traffic while some may be carrying GPRS data. Moreover, GPRS allows for dynamic allocation of resources, such that a given timeslot may be used for standard voice traffic and subsequently for GPRS data traffic, depending on the relative traffic demands. Therefore, the RF design or frequency planning required for GPRS is essentially the same as that required for GSM.

While GPRS does use the same basic structure as GSM, the introduction of GPRS means the introduction of a number of new logical channel types and new channel coding schemes to be applied to those logical channels. When a given timeslot is used to carry GPRS-related data traffic or control signaling then it is known as a *Packet Data Channel* (PDCH).

Similar to GSM, GPRS requires a number of control channels. To begin, there is the *Packet Common Control Channel* (PCCCH), which, like the CCCH in GSM, comprises a number of logical channels. The logical channels of the PCCCH include:

- *Packet Random Access Channel (PRACH)*. It is applicable only in the uplink and is used by an MS to initiate transfer of packet signaling or data.

- *Packet Paging Channel (PPCH)*. It is applicable only in the downlink and is used by the network to page an MS prior to downlink packet transfer.

- *Packet Access Grant Channel (PAGCH)*. It is applicable only in the downlink and is used by the network to assign resources to the MS prior to packet transfer.

- *Packet Notification Channel (PNCH)*. It is used for *Point-To-Multipoint–Multicast* (PTM-M) notifications to a group of MSs.

The PCCCH must be allocated to an RF resource (i.e., a different timeslot) different from the CCCH. The PCCCH, however, is optional. If it is omitted, then the necessary GPRS-related functions are supported on the CCCH.

Similar to the BCCH in GSM, GPRS includes a *Packet Broadcast Control Channel* (PBCCH). This is used to broadcast GPRS-specific system information.

4.5.4 GPRS network architecture

GPRS is effectively a packet data network overlaid on the GSM network. It provides packet data channels on the air interface and a packet data switching and transport network that is largely separate from the standard GSM switching and transport network.

Figure 4.20 shows the GPRS network architecture. One can see that there are a number of new network elements and interface as compared to Fig. 4.16. In particular, we find the *Packet Control Unit* (PCU), the *Serving GPRS Support Node* (SGSN), the *Gateway GPRS Support Node* (GGSN), and the *Charging Gateway Function* (CGF).

The PCU is a logical network element that is responsible for a number of GPRS-related functions such as air interface access control, packet scheduling on the air interface, packet assembly, and reassembly. Strictly speaking, the PCU can be placed at the BTS, at the BSC, or at the SGSN. Logically, the PCU is considered a part of the BSC and in real implementations one finds the PCU physically integrated with the BSC.

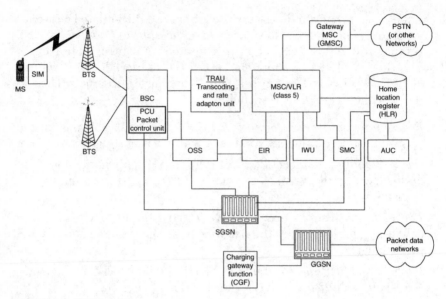

Figure 4.20 GPRS architecture.

The SGSN is analogous to the MSC/VLR in the circuit-switched domain. Just as the MSC/VLR performs a range of functions in the circuit-switched domain, the SGSN performs the equivalent functions in the packet-switched domain. These include mobility management, security, and access control functions.

The service area of an SGSN is divided into *routing areas* (RAs) that are analogous to *location areas* in the circuit-switched domain. When a GPRS MS moves from one RA to another, it performs a *Routing Area Update*, which is similar to a Location Update in the circuit-switched domain. One difference, however, is that an MS may perform a routing area update during an ongoing data session, which in GPRS terms is known as a *Packet Data Protocol* (PDP) *Context*. In contrast, for an MS involved in a circuit-switched call, a change of location area does not cause a location update until after the call is finished.

A given SGSN may serve multiple BSCs, whereas a given BSC interfaces with only one SGSN.

A GGSN is the point of interface with external packet data networks (e.g., the Internet). Thus, user data enters and leaves the PLMN via a GGSN. A given SGSN may interface with one or more GGSNs and the interface between an SGSN and GGSN is known as the *Gn interface*. This is an IP-based interface used to carry signaling and user data. The Gn interface uses the *GPRS Tunneling Protocol* (GTP), which tunnels user data through the IP backbone network between the SGSN and GGSN.

4.5.5 GPRS attach

GPRS functionality in an MS may be activated either when the MS itself is powered on, or perhaps when the browser is activated. Whatever the reason for the initiation of GPRS functionality within the MS, the MS must attach to the GPRS network so that the GPRS network (and specifically the serving SGSN) knows that the MS is available for packet traffic. In the terms used in GPRS specifications, the MS moves from idle state (not attached to the GPRS network) to ready state (attached to the GPRS network and in a position to initiate a PDP context). When in ready state, the MS can send and receive packets. There is also a standby state, which the MS enters after a time-out in the ready state. If, for example, the MS attaches to the GPRS network but does not initiate a session, it will then remain attached to the network, but move to standby state after a time-out. Figure 4.21 shows the simple case of an MS performing a GPRS attach.

4.5.6 Establishing a PDP context

The transfer of packet data is through the establishment of a PDP context, which is effectively a data session. Normally, such a context is initiated by the MS, as would happen, for example, when a browser on the MS is activated and the subscriber's home page is retrieved from the Internet. When an MS, or the network, initiates a PDP context, the MS moves from the standby state to the ready state. The initiation of a PDP context is illustrated in Fig. 4.22.

4.6 Enhanced Data Rates for Global Evolution (EDGE)

EDGE once stood for the term Enhanced Data Rates for GSM Evolution. The basic goal with EDGE is to enhance the data throughput capabilities of a GSM/GPRS network. This is done primarily by changing the air interface modulation scheme from *Gaussian Minimum Shift Keying* (GMSK), as used in GSM, to *8 Phase Shift Keying* (8-PSK). The result is that EDGE can theoretically support speeds of up to 384 kbps. Thus, it is clearly more advanced than GPRS, but still does not meet the requirements for a true 3G system (which should support speeds of up to 2 Mbps).

To deploy EDGE instead does not require new spectrum and does not require as drastic changes to the network.

The network architecture for EDGE uses GPRS as the foundation. EDGE uses the same network elements, the same interfaces, the same protocols, and the same procedures as those for GPRS. There are some minor differences in the network primarily on the RAN.

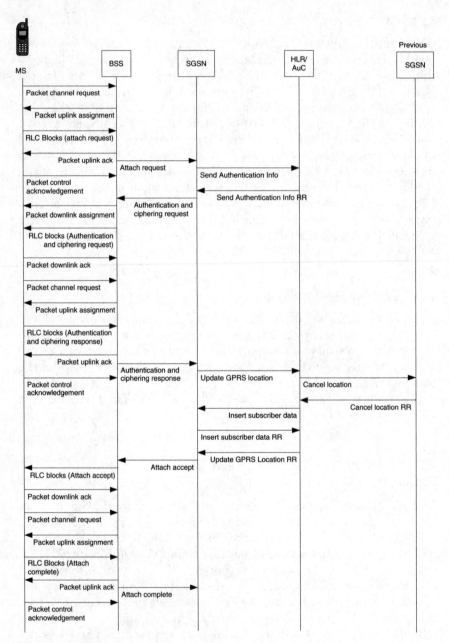

Figure 4.21 GPRS attach.

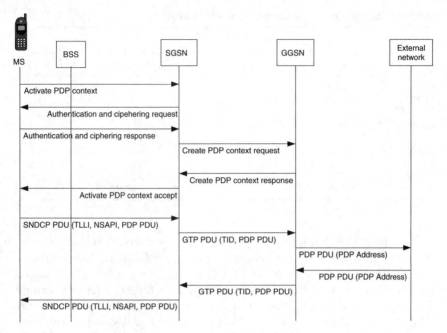

Figure 4.22 PDP context activation.

EDGE uses the same 200 kHz channels and eight-timeslot structure as used for GSM and GPRS. With EDGE, however, 8 PSK modulation is introduced in addition to the 0.3 GMSK used in GSM.

The objective with EDGE is to offer higher bandwidth efficiency so that we can squeeze more user data from the same 200 kHz channel. This higher bandwidth efficiency is achieved through 8-PSK. In general, PSK involves a phase change of the carrier signal according to the incoming bit stream.

However, the increased throughput comes at a price from the RAN and that is 8-PSK is more sensitive to noise than GMSK. The direct result of this is that if a BTS supports both GMSK and 8-PSK modulation and has the same output power for both, then the cell footprint is smaller for 8-PSK than for GMSK.

Recognizing this limitation, however, the specifications for EDGE are such that both the coding scheme and modulation scheme can be changed in response to RF conditions. Thus, as a user moves toward the edge of a cell, the effect of lower signal to noise will mean that the network can reduce the user's throughput, either by changing the modulation scheme to GMSK or by changing the coding scheme to include greater error detection. All that the user will notice is somewhat slower throughput.

TABLE 4.6   Modulation and Coding Schemes for EGPRS

Scheme	Modulation	RLC Blocks per radio block (20 ms)	Input data payload (bits)	Date rate (kbps)
MCS-1	GMSK	1	176	8.8
MCS-2	GMSK	1	224	11.2
MCS-3	GMSK	1	296	14.8
MCS-4	GMSK	1	352	17.6
MCS-5	8-PSK	1	448	22.4
MCS-6	8-PSK	1	592	29.6
MCS-7	8-PSK	2	2×448	44.8
MCS-8	8-PSK	2	2×544	54.4
MCS-9	8-PSK	2	2×592	59.2

Therefore, EDGE introduces a number of new channel coding schemes in addition to the coding schemes that exist for GSM voice and GPRS. For packet data services in an EDGE network, we refer to *Enhanced GPRS* (EGPRS) and the new coding schemes for EGPRS are termed *Modulation and Coding Scheme-1* to *Modulation and Coding Scheme-9* (MCS-1 to MCS-9). The reason they are not just called coding schemes is that, for MCS-1 to MCS-4, GMSK modulation is used, whereas 8-PSK modulation is used for MCS-5 to MCS-9. Table 4.6 shows the modulation scheme and data rate applicable to each MCS.

The channel types applicable to EGPRS are the same as those applicable to GPRS, that is, we have a number of PDCHs that carry PCCCH, PBCCH, PDTCHs, and the like. In fact, these channels are shared among GPRS and EGPRS users. Thus, both GPRS users and EGPRS users can be multiplexed on a given PDTCH. Of course, during those radio blocks that the PDTCH is used by an EGPRS user, then the modulation may be either GMSK or 8-PSK, whereas it must be GMSK when used by a GPRS user.

Similar to the manner in which the network controls the coding scheme to be used by a GPRS user, the network also controls the MCS to be used by an EGPRS user in both the uplink and downlink.

4.7   Universal Mobile Telecommunications Service (UMTS)

UMTS is the next stage in the evolution of GSM to support 3G capabilities. The following section will attempt to present a high-level overview of UMTS, the main components being most applicable to the integration of mobility with fixed wireless access. A more detailed description of UMTS can be found in the references for this chapter. However, it is important to keep in mind when migrating from GSM/GPRS/EDGE to UMTS that the RAN is fundamentally different with little or no commonality.

UMTS includes two of the air interface proposals submitted to the International Telecommunications Union (ITU) as proposed solutions to meet the requirements laid down for *International Mobile Telephony 2000* (IMT-2000). Both solutions use *Direct Sequence Wideband CDMA* (DS-WCDMA), however, one uses the Frequency Division Duplex (FDD) method and the other uses the Time Division Duplex (TDD) method.

With the FDD option there are paired 5 MHz carriers in the uplink and downlink as follows: Uplink, 1920 MHz to 1980 MHz; Downlink, 2110 MHz to 2170 MHz. Thus, for the FDD mode of operation, there is a separation of 190 MHz between uplink and downlink.

For the TDD option, a number of frequencies have been defined, including 1900 MHz to 1920 MHz and 2010 MHz to 2025 MHz. Of course, with TDD, a given carrier is used in both the uplink and the downlink so that there is no separation.

From a network architecture perspective, UMTS borrows heavily from the established core network architecture of GSM. In fact, many of the network elements used in GSM are reused (with some enhancements) in UMTS. This commonality means that a given MSC, HLR, SGSN, or GGSN can be upgraded to support UMTS and GSM simultaneously. The radio access, however, is significantly different from that of GSM, GPRS, and EDGE. In UMTS the RAN is known as the *UMTS Terrestrial Radio Access Network* (UTRAN). The components that make up the UTRAN are significantly different from the corresponding elements in the GSM architecture. Therefore, reuse of existing GSM base stations and Base Station Controllers is limited.

The reuse of portions of the GSM BTS is dependant on the infrastructure vendor used. Thus, depending on the vendor used, it is possible to remove some or all of the GSM transceivers from a base station and replace them with some number of UMTS transceivers. However, for other vendors, a completely new base station is needed.

A similar situation applies to BSCs. For most vendors, the technology of a UMTS *Radio Network Controller* (RNC) is so different from that of a GSM BSC that the BSC cannot be upgraded to act simultaneously as a GSM BSC and a UMTS RNC. There are cases, however, where a BSC can be upgraded to simultaneously support both GSM and UMTS, but that situation is less common.

UMTS for the RAN uses a different modulation scheme than GSM but similar to EDGE in that it uses QPSK for both the uplink and downlink. However, the channel bandwidth is 5 MHz as compared to 200 kHz.

Like its counterpart, CDMA2000, UMTS uses a series of codes for spreading the signal. The downlink codes are *gold codes* similar to the long scrambling codes used in the uplink. The codes are separated into 512 groups. Each group contains one primary code and 15 secondary codes. Thus, there are 512 primary codes and 7680 secondary codes, for a total of 8192 downlink codes.

A cell is allocated one primary code, which, of course, has 15 secondary codes associated with it. A given base station will use its primary code for transmission of channels that need to be heard by all terminals in the cell. All transmissions from the base station can simply use the cell's primary code. Various channelization codes are used to separate the various transmissions (i.e., physical channels) within the cell.

4.7.1 Channel types

At the physical layer, the *User Equipment* (UE), UMTS subscriber, and the network communicate via a number of physical channels. In general, there are two types of transport (physical) channels. These are common transport channels and dedicated transport channels. Common transport channels may be applicable either to all users in a cell or to one or more specific users.

The common transport channels are:

- *Random Access Channel (RACH).* It is used in the uplink when a user wishes to gain access to the network. The RACH is available to all users in the cell.

- *The Broadcast Channel (BCH).* It is used in the downlink to transmit system information over the entire coverage area of a cell.

- *The Paging Channel (PCH).* It is used in the downlink to page a given UE when the network wishes to initiate communication with a user.

- *The Forward Access Channel (FACH).* It is used to send downlink control information to one or more users in a cell.

- *The Uplink Common Packet Channel (CPCH).* It is similar to the RACH but can last for several frames. Thus, it allows for a greater amount of data to be sent than is allowed by the RACH.

- *The Downlink Shared Channel (DSCH).* It is used to carry dedicated user data or control signaling to one or more users in a cell. It is similar to the FACH, but does not have to be transmitted over the entire cell area.

There is only a single dedicated transport channel type, the *Dedicated Channel* (DCH). The DCH carries user data and is specific to a single user. The DCH is designed for large amounts of data or for extended data sessions. The data rate on a DCH can vary on a frame-by-frame basis.

The transport channels are mapped to specific physical channels for the UMTS air interface. In general, a physical channel is identified by a specific frequency, scrambling and channelization codes, duration, and in the uplink, phase. In addition to the physical channels that are mapped to/from transport channels, there are a number of physical channels that exist only for the correct operation of the physical layer. Such channels are not visible to higher layers.

The physical channels are:

- The *Synchronization Channel (SCH)*. It is transmitted by the base station and is used by a UE during the cell search procedure. The SCH is transmitted in conjunction with the *Primary Common Control Physical Channel* (primary CCPCH) described below.

- The *Common Pilot Channel (CPICH)*. This is a channel always transmitted by the base station and scrambled with the cell-specific primary scrambling code. It uses a fixed spreading factor of 256, which equates to 30 kbps on the air interface. An important function of the CPICH is in measurements by the terminal for handover or cell reselection as the measurements made by the terminal are based on the reception of the CPICH.

- The *Primary Common Control Physical Channel (Primary CCPCH)*. It is used on the downlink to carry the BCH transport channel.

- The *Secondary Common Control Physical Channel (Secondary CCPCH)*. It is used on the downlink to carry two common transport channels—the FACH and the PCH. The FACH and the PCH can share a single secondary CCPCH or each can have a secondary CCPCH of its own. The secondary CCPCH carrying the PCH must be transmitted over the whole cell area, which applies regardless of whether the physical channel carries just the PCH or both PCH and FACH.

- The *Physical Random Access Channel (PRACH)*. It is used in the uplink to carry the RACH transport channel.

- The *Physical Common Packet Channel (PCPCH)*. It is used in the uplink to carry the uplink CPCH transport channel.

- The *Physical Downlink Shared Channel (PDSCH)*. It is used in the downlink to carry the DSCH transport channel. Because the DSCH transport channel can be shared among several users, the PDSCH has a structure that allows it to be shared among users.

- The *Indicator Channels*. These include the AICH, AP-AICH, and CD/CA-ICH already mentioned. In addition, there is the *Paging Indicators Channel* (PICH). The purpose of the PICH is to let a given terminal know when it might expect a paging message on the PCH (carried on the secondary PCPCH).

- The *DCH transport channel*. It is mapped to the two physical channels, DPDCH and DPCCH. The DPDCH carries the actual user data and can have a variable spreading factor, while the DPCCH carries control information.

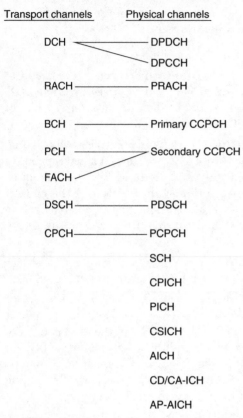

Figure 4.23 UMTS transport and physical channel mapping.

The mapping between the transport channels and physical channel is shown in Fig. 4.23.

The logical channels are mapped to transport channels, which in turn, are mapped to physical channels. Logical channels relate to what information is being transmitted, while transport channels relate largely to the manner in which the information is transmitted. There are basically two groups of logical channels, control and traffic channels with their relationship shown in Fig. 4.24.

- The *Broadcast Control Channel (BCCH)*. It is used for downlink transmission of system information.

- The *Paging Control Channel (PCCH)*. This is used for paging of an MS across one or more cells.

- The *Common Control Channel (CCCH)*. It is used in the uplink by terminals that wish to access the network but do not already have any

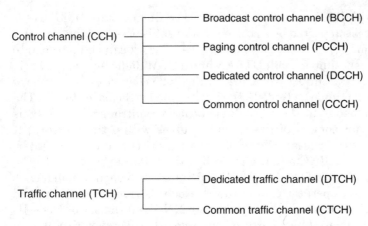

Figure 4.24 Logical channels.

connection with the network. The CCCH can be used in the downlink to respond to such access attempts.

- The *Dedicated Control Channel (DCCH)*. It is a bidirectional point-to-point control channel between the MS and the network for sending control information. WCDMA also defines the *Shared Channel Control Channel*, but that channel is used only in TDD mode.

- *Logical Traffic Channels*. There are two types of logical traffic channels. The *Dedicated Traffic Channel* (DTCH) is a point-to-point channel, dedicated to one UE, for the transfer of user data. DTCHs apply to the uplink and the downlink. The *Common Traffic Channel (CTCH)* is a point-to-multipoint unidirectional channel for transfer of user data to all UEs or just to a single UE. The CTCH exists in the downlink only.

WCDMA is designed to offer great flexibility in transmission of user data across the air interface. For example, data rates can change on a frame-by-frame basis (every 10 ms). Moreover, it is possible to support mix and match of different types of services. For example, a subscriber may be sending and receiving packet data while also involved in a voice call. When sending information over the air interface, physical controls channels are used in combination with physical data channels. While the physical data channels carry the user information, the physical control channels carry information to support the correct interpretation of the data carried on the corresponding DPDCH frame, power control commands, and feedback indicators.

4.7.2 UTRAN architecture

Mobile communication networks are traditionally defined in terms of their RAN and core network architecture. UMTS also uses this division.

In UMTS the RAN is known as the UMTS Terrestrial Radio Access Network. It is supported by a core network that is based on the core network used for GSM. In fact, the GSM core network can be upgraded to simultaneously support both UTRAN and a GSM Radio Access Network.

The UTRAN architecture shown in Fig. 4.25 comprises two types of nodes, the RNC and the Node B, the latter which is the base station. The RNC is analogous to the GSM Base Station Controller. The RNC is responsible for control of the radio resources within the network. It interfaces with one or more base stations known as Node Bs. The interface between the RNC and the Node B is the Iub interface. Unlike the equivalent Abis interface is GSM, the Iub interface is open, which means that a network operator could acquire Node Bs from one vendor and RNCs from another vendor. Together, an RNC and the set of Node Bs that it supports are known as a *Radio Network Subsystem* (RNS).

UTRAN also has an interface between the RNCs unlike in GSM. This is known as the Iur interface and its purpose is to support inter-RNC mobility and soft handover between Node Bs connected to different RNCs.

The user device is known as the User Equipment. It comprises the ME and the *UTMTS Subscriber Identity Module* (USIM). UTRAN communicates with the UE over the Uu interface. The Uu interface is none other than the WCDMA air interface as compared to the Um in GSM.

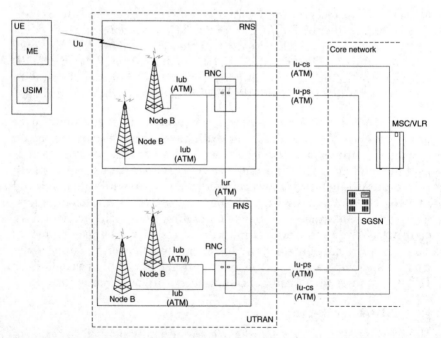

Figure 4.25 UTRAN architecture.

UTRAN communicates with the core network over the Iu interface. In 3GPP Release 1999, the Iu interface has two components: the Iu-CS interface, which supports circuit-switched services, and the Iu-PS interface, which supports packet-switched services. The Iu-CS interface connects the RNC to an MSC and is similar to the GSM A-interface. The Iu-PS interface connects the RNC to an SGSN and is analogous to the GPRS Gb interface. All the UTRAN interfaces with the core network are via ATM.

The RNC shown in Fig. 4.25 that controls Node B is known as the *Controlling RNC* (CRNC). The CRNC is responsible for the management of radio resources supported by the Node B that it supports.

For a given connection between the UE and the core network, there is one RNC in control. This is called the *Serving RNC* (SRNC). For the user in question, the SRNC controls the radio resources that the UE is using. In addition, the SRNC terminates the Iu interface to/from the core network for the services being used by the UE. In many cases, though not all, the SRNC is also the CRNC for a Node B that is serving the user.

UTRAN also supports soft handover, like CDMA2000, which may occur between Node Bs controlled by different RNCs. During and after soft handover between RNCs, a UE could be communicating with a Node B that is controlled by an RNC that is not the SRNC. Such an RNC is termed a *Drift RNC* (DRNC). The DRNC does not perform any processing of user data (beyond what is required for correct operation of the physical layer). Rather, data to/from the UE is controlled by the SRNC and is passed transparently through the DRNC.

As a UE moves further and further away from any Node B controlled by the SRNC, it will become clear that it is no longer appropriate for the same RNC to continue to act as the SRNC. In that case, UTRAN may make the decision to hand the control of the connection over to another RNC. This is known as *Serving RNS* (SRNS) relocation. This action is invoked under the control of algorithms within the SRNC.

4.7.3 Call flow diagram

The overall call flow for a UMTS basic speech conversation is shown in Fig. 4.26.

4.7.4 UMTS packet data

UMTS packet data services use largely the same mechanisms as used for GPRS data with the distinguishing difference being the increased user data rates supported.

Another key enhancement or difference is that the Gb interface of GPRS (between the SGSN and BSC) is replaced by the Iu-PS interface, which uses RANAP as the application protocol. This change includes the fact that IP over ATM is used between the SGSN and RNC. Thus, there

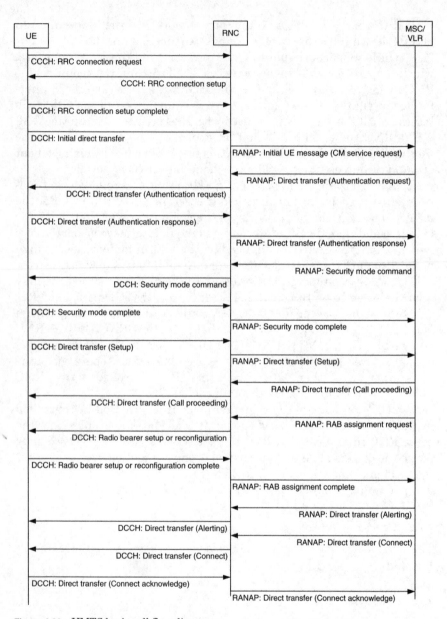

Figure 4.26 UMTS basic call flow diagram.

is an IP network from GGSN to SGSN to RNC. Consequently, the GTP-U tunnel can be relayed from the GGSN through the SGSN to the RNC, rather than terminating at the SGSN. The GTP-C tunnel, however, does terminate at the SGSN, since the application protocol between

RNC and SGSN is RANAP rather than GTP. The establishment of the tunnel is still under the control of the SGSN.

Therefore, packet data services are established in UMTS in largely the same manner as in GPRS—through the activation of a PDP context with an *access point name* (APN), quality of service criteria, and the like. One significant difference between UMTS and standard GPRS, however, involves SRNS relocation. Because of the fact that the GTP-U tunnel terminates at the RNC rather than the SGSN, relocation of the UE to another RNC may require the buffering of packets at the first RNC and subsequent relay of those packets to the second RNC once relocation has taken place. That relay occurs via the SGSN. In the case that the two RNCs are connected to two different SGSNs, the path for buffered packets is from RNC1 to SGSN1 to RNC2 to SGSN2.

From an air interface perspective, UMTS provides greater flexibility than GPRS in terms of how resources are allocated for packet data traffic. Not only does UMTS offer a greater range of speeds, the WCDMA air interface has a selection of different channel types that can be used for packet data. In the uplink, the RACH, CPCH, and DCH are available. In the downlink, the FACH and DSCH are available.

4.7.5 UMTS core network

3GPP introduces a significant enhancement to the core network architecture as it applies to the CS domain. Basically, the MSC is broken into constituent parts and is allowed to be deployed in a distributed manner as shown in Fig. 4.27. Specifically, the MSC is divided into an MSC server and a *media gateway* (MGW). The MSC server contains all of the mobility

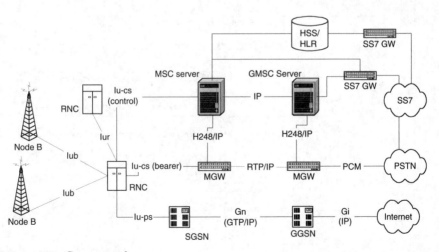

Figure 4.27 Core network.

management and call control logic that would be contained in a standard MSC. It does not, however, reside in the media path. Rather, the media path is via one or more MGWs that establish, manipulate, and release media streams (e.g., voice streams) under the control of the MSC server.

Control signaling for circuit-switched calls is between the RNC and the MSC server. The media path for circuit-switched calls is between the

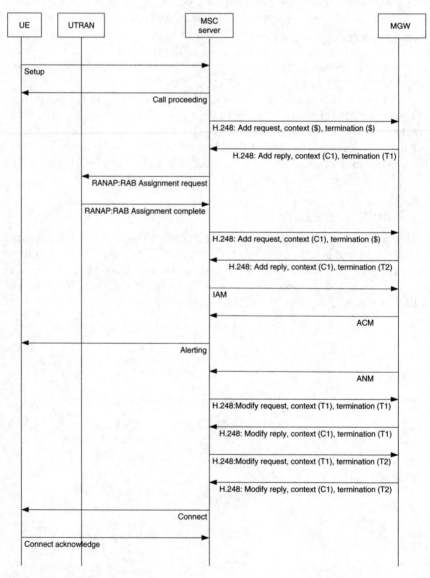

Figure 4.28 Call flow diagram.

RNC and the MGW. Typically, an MGW takes calls from the RNC and routes those calls toward their destination over a packet backbone.

The control protocol between the MSC server or GMSC server and the MGW is the ITU H.248 protocol. This protocol is also known as *MEGACO*. The call control protocol between the MSC server and the GMSC server can be any suitable call control protocol.

Figure 4.28 shows an example of a voice call establishment using this architecture. For the sake of brevity, the messages are limited to those that relate to call establishment as seen from the core network. Thus, the RRC protocol messages between the UE and UTRAN have been omitted.

Note that the distributed architecture just described does not rely on the fact that there is a WCDMA-based UTRAN access. The core network architecture could just as well apply to a standard GSM-based access, with BSCs instead of RNCs. In fact, as the distributed switching architecture is being deployed, it is likely that many deployments will simultaneously support both UTRAN and GSM access networks.

References

Bates, R. J., and D. Gregory, *Voice and Data Communications Handbook*, Signature Edition, McGraw Hill, New York, 1998.

Brewster, R. L., *Telecommunications Technology*, Wiley, New York, 1986.

DeRose, J., *The Wireless Data Handbook*, Quantum Publishing Inc., Mendocino, CA, 1994.

Harte, L., M. Hoenig, D. McLaughlin, and R. Kikta, *CDMA IS-95 for Cellular and PCS*, McGraw-Hill, New York, 1996.

Held, G., *Voice & Data Interworking*, 2nd ed., McGraw-Hill, New York, 2000.

Homa, H., and A. Toskala, *WCDMA for UMTS*, Wiley, 2000.

Lee, W.C.Y., *Mobile Cellular Telecommunications Systems*, 2d ed., McGraw-Hill, New York, 1996.

Molisch, A. F., *Wideband Wireless Digital Communications,* Prentice Hall, Upper Saddle River, New Jersey, 2001.

Mouly, M., and M.-B. Pautet, *The GSM System for Mobile Communications*, Mouly Pautet, France, 1992.

Muratore, F., *UMTS Mobile Communications for the Future*, Wiley, Sussex England, 2000.

Prasad, R., W. Mohr, and W. Konhauser, *Third Generation Mobile Communication Systems*, Artech House, Boston, 2000.

Rappaport, T., *Wireless Communications Principals and Practices*, IEEE, New York, 1996.

Shneyderman, A., and A. Casti, *MobileVPN*, Wiley, New York, 2003.

Smith, C., *LMDS*, McGraw-Hill, New York, 2000.

Smith,C., *Practical Cellular and PCS Design*, McGraw-Hill, New York, 1997.

Smith, C., *Wireless Telecom FAQ*, McGraw-Hill, New York, 2000.

Smith,C., and D. Collins, *3G Wireless Networks*, McGraw-Hill, New York, 2001.

Smith,C., and C. Gervelis, *Wireless Network Performance Handbook*, McGraw-Hill, New York, 2003.

5

802.11

Wi-Fi is a wireless LAN based on the 802.11 Standards. The prevalence of Wi-Fi is evident through advertisements and is now a standard feature for laptops, computers, and PDAs. However, Wi-Fi is also migrating into the wireless mobility arena as it enhances the customer's experience with wireless data.

There are a number of standards that make up 802.11 with the most popular being referred to as Wi-Fi. This chapter's objective is to provide an overview of the various flavors of 802.11. In addition, this chapter will introduce and better define the interaction of 802.11 and wireless mobility.

It is not intended to present a detailed definition of the 802.11 specifications since they can be found in the appropriate Institute of Electrical and Electronics Engineers documents, www.IEEE.org. Instead, this chapter will attempt to make sense out of the plethora of information and options available for a mobile wireless operator to integrate 802.11 into the wireless data and overall service offerings.

5.1 Wireless LANs

Wi-Fi is a certification provided by the *Wireless Ethernet Compatibility Alliance* (WECA).

Components with 802.11 Wi-Fi certification are guaranteed to be interoperable. The term Wi-Fi is becoming synonymous with wireless LAN. The guarantee of interoperability with numerous vendors makes it the implementation of choice for Public Hotspot Networks.

Wireless Local Area Networks (WLAN) have been around since the 1980s. Earlier versions employed proprietary protocols. This of course precluded interoperability. In other words, when implementing a wireless LAN it was necessary to use equipment from a single manufacturer. This made interoperability problematic and also did not provide for the economies of scale afforded by a more universally adopted standard for WLAN.

The 802.11 Standard is part of a family of IEEE Standards devoted to defining Local and Wide Area Networks. These standards deal with the physical and data link layers defined by the *International Standards Organization Open Systems Interconnect* (ISO-OSI) reference model (ISO/IEC 7498-1:1994.). For a detailed view of this family of standards see IEEE STD 802-2001. A short overview of pertinent parts follows.

5.2   802 Standard

This family of standards deals with the physical and data link layers as defined by the ISO-OSI basic reference model (ISO/IEC 7498-1:1994). The access standards define seven types of medium access technologies and associated physical media, each appropriate for particular applications or system objectives. Other types are under investigation.

The OSI/RM is used for the following reasons:

- The OSI/RM provides a common vehicle for understanding and communicating the various components and interrelationships of the standards.
- The OSI/RM defines terms.
- The OSI/RM provides a convenient framework to aid in the development and enhancement of the standards.
- The use of the OSI/RM facilitates a higher degree of interoperability than might otherwise be available.

The 802 Standards define the physical and link layers for various media access controls to be transparent to upper levels of the OSI reference model. This allows for the interconnection of various LANs without regard to their physical characteristics such as being wireless.

The standards defining the access technologies are shown in Table 5.1.

5.3   802.11 Objectives

The 802.11 Standard addresses the following:

- Describes the functions and services required by an IEEE 802.11 compliant device
- Defines the *media access control* (MAC) procedures to support the asynchronous *MAC service data unit* (MSDU) delivery services

TABLE 5.1   802 Standards

Standard	Topic addressed
802	Overview and architecture of the family of standards
802.1b	LAN/MAN remote management
802.1D	MAC Bridging (interconnecting of 802 LANS)
802.1E	System loading and QOS
802.1F	Procedures for management information of 802 LANs
802.1G	Remote MAC bridging (non-LAN technologies)
802.2	Logical link control
802.3	CSMA/CD "Ethernet" physical layer
802.4	Token bus access method and physical layer
802.5	Token ring access method and physical layer
802.6	Dist. queue dual bus access method
802.9	Integrated services LAN interface MAC and physical layers
802.11	Wireless LAN MAC and physical layers (Wi-Fi)
802.12	Demand priority access method
802.7	Broadband LAN's
802.14	Cable TV protocols
802.15	Wireless PAN
802.16	Wireless LAN/MAN
802.20	3G wireless data

- Defines several PHY signaling techniques and interface functions that are controlled by the IEEE 802.11 MAC

- Enables the operation of an IEEE 802.11 conformant device within a wireless LAN

- Describes the requirements for privacy, security, and authentication

Other members of the 802 family of standards that are of immediate interest follow in Table 5.2.

Other standards of interest include 802.3, which describes the Ethernet that is employed in many wired LANs. Another wireless standard that addresses Wide Area Networks (WAN) is 802.16.

This standard covers the frequency range of 2 GHz to 66 GHz. It includes license-exempt bands similar to those used by Wi-Fi (ISM band and UN-II band). This specification also includes the LMDS band which is a license band.

It will be interesting to see if the advent of this specification breathes new life into the LMDS/MMDS world. When this spectrum was originally auctioned no standard protocols were available. This made equipment selection more difficult. More importantly, equipment pricing was substantially higher due to lack of economies of scale.

Time will tell if 802.16 is adopted as widely as Wi-Fi. Early signs are good with industry giant Intel announcing they will have 802.16 products some time in 2004. Due to the fact that much of this spectrum is licensed

TABLE 5.2   General Standards Covering All Physical
and Medium Access Standards

Institute of Electrical and Electronics Engineers Standard	Topic addressed
802	Overview
802.1B	Remote management
802.2	Logical layer
802.1	Bridging
Specific LAN/WAN standards	
802.3	CSMA/CD "Ethernet"
802.10	Security
802.11	Wireless LAN MAC "Wi-Fi"
802.16	Fixed wireless access broadband systems includes LMDS spectrum

(available for use only to the license holder), interference and congestion become a controllable issue for the operator. In addition, the use of higher output power and high-gain antennae will allow range extension for 802.16 systems. Chapter 6 is devoted to an in-depth discussion of 802.16.

After years of development and research the 802.11b Standard was published by the Institute of Electrical and Electronics Engineers in 1999. The existence of a standard allowed for the development of devices by multiple manufacturers and thus resulted in far less expensive solutions. The adoption of the standard by giants such as Intel and Cisco almost guaranteed wide availability.

Once this was recognized, a number of companies began deploying "public" WLANs in Hot Spots at hotels, airports, coffee shops, and even McDonalds. This, along with the use of WLANs in the home and at work, is beginning to create a mass demand for 802.11 or Wi-Fi devices. The presence of a Wi-Fi adapter in today's laptops is copying the history of the NIC card and modem. While these items were once options they are now an expected standard feature in off-the-shelf computers today.

5.4   Wi-Fi

There are three main variants of the 802.11 Standard that are most common in today's WLAN. They are 802.11b, 802.11g, and 802.11a. Details of the differences in these standards are addressed later in the chapter. The letter designator does not imply the historical availability of the standard. Two of the standards 802.11b and 802.11g have interoperable characteristics. 802.11a operates in a different frequency band and while it may coexist in a WLAN its components will not interoperate with 802.11b and 802.11g.

5.4.1 802.11b

This is the most widespread variant of the standards. It's the one you would most likely find in Hot Spots. This standard was published in 1999 and has been widely adapted by manufacturers of infrastructure such as access points, routers, and bridges. It is also widely adapted by vendors of interface devices for laptops, desktops, and PDAs.

Standard 802.11b operates in the Industrial, Scientific, and Medical band (ISM) at 2.4 GHz. Standard 802.11 specifies data rates up to 11 Mbps. The standard specifies *Direct Sequence Spread Spectrum* (DSS) and several modulation schemes, CCK and PBCC.

5.4.2 802.11g

This standard provides higher data rates (up to 54 Mbps). The g standard employs DSS/FSSS and *Orthogonal Frequency Division Multiplexing* (OFDM). Standard 802.11g is backward compatible with 802.11b. By this we mean that any 802.11g device must be able to coexist with 802.11b devices. An 802.11g-equipped laptop must work in an 802.11b AP coverage area. In addition, if an 802.11b laptop comes into an 802.11g AP coverage area, the 802.11g AP must be able to serve the device.

This backward compatibility helps ease the transition of introduction of 802.11g components into the system by not obsolescing 802.11b devices. A drawback to this is the fact that when an 802.11b device is detected in the 802.11g serving area, additional overhead is introduced. This overhead diminishes actual throughput by as much as 25 percent.

5.4.3 802.11a

802.11a is a higher speed WLAN solution operating in the UNII band at 5 GHz. The g standard uses OFDM. OFDM is discussed in much greater detail in Chap. 6. Standard 802.11a can operate at 54 Mbps. While 802.11a and 802.11b are not compatible, it is not unusual to use them both in an Enterprise Network. The majority of users may be employing 802.11b while power users are assigned to 802.11a. These networks would be overlapping and not interoperating. Although not required by any standard, there are access points available that offer both 802.11b and 802.11a.

TABLE 5.3   802.11 Comparison

Function	802.11b	802.11g	802.11a
Maximum data rate	11 Mbps	54 Mbps	54 Mbps
Number of nonoverlapping channels	3	3	11
Frequency allocation	2.4 Ghz ISM Band	2.4 Ghz ISM Band	5 Ghz UNII band
Modulation/coding schemes	CCK, HR/DSSS, HR/DSS/short HR/DSSS/PBCC	OFDM	OFDM
Spread spectrum type	DSS	FSS	FSS

5.4.4   Comparison of 802.11a, 802.11b, and 802.11g

The attributes of each of these standards was discussed earlier. Each has its place in WLAN architecture and indeed all the three can be used to accomplish your wireless network requirements. Table 5.3 compares some of the main attributes of each.

5.5   Frequency Allocation

The 802.11 Standards address two frequency ranges. 802.11b and 802.11g operate in the 2.4 GHz band (ISM) while 802.11a operates in the 5 GHz band (UN-II). Each has a set of channel numbers associated to it in the appropriate band. The availability of specific channels varies among regions. For example, in the United States there are 11 channels available to 802.11b and 802.11g. The reality (discussed later in the chapter) is that only three of the channels can be used together without overlap.

5.5.1   802.11b and 802.11g

Table 5.4 shows channel allocation within the ISM band for 802.11b, g services for a number of regions around the world. It is important to note that the spectrum or channel allocations are not uniform throughout the world.

5.5.2   802.11a

The 802.11a frequency and corresponding channel number are shown in Table 5.5. These channels correspond to the 5 GHz UN-II band.

5.6   Modulation and Coding Schemes

There are several coding schemes used by the Wi-Fi variants of 802.11. While Wi-Fi is a CDMA technology, the modulation and coding schemes

TABLE 5.4   802.11b, g Channel Chart

Channel	Frequency GHz	US	Europe	Spain	France
1	2412	X	X		
2	2417	X	X		
3	2422	X	X		
4	2427	X	X		
5	2432	X	X		
6	2437	X	X		
7	2442	X	X		
8	2447	X	X		
9	2452	X	X	X	X
10	2447	X	X	X	X
11	2462	X	X		X
12	2467		X		X
13	2472		X		
14	2484				

are significantly different from those of 2.5/3G technologies used by mobile wireless systems.

5.6.1   802.11b

In order to develop higher data rates for 802.11b, several modulation schemes were added to the standard.

Complimentary code keying (CCK). The addition of this modulation scheme enables higher data rates of 5.5 Mbps and 11 Mbps. The higher data rate capability is known as *High Rate Direct Sequence Spread Spectrum* (HR/DSSS). The same preamble and header are used as the earlier basic rate DSSS method. This allows the HR/DSSS and basic DSSS to coexist in the same BSS.

TABLE 5.5   802.11a Channels

Band	Channel number	Center frequency
Lower	36	5180
	40	5200
	44	5220
	48	5240
Center	52	5260
	56	5280
	60	5300
	64	5320
Upper	149	5745
	153	5765
	157	5785
	161	5805

The standard also offers optional features that can be used to increase data throughput. These features will allow vendors to differentiate their product offerings and include HR/DSSS/short—essentially HR/DSSS with a short preamble and HR/DSSS/PBCC (HR/DSSS Packet Binary Convolutional Coding). This optional implementation can coexist under limited conditions. For example, they could be implemented on different channels.

5.6.2   802.11g and 8021.11a

Both of these standards incorporate the modulation scheme of OFDM. This topic is addressed in depth in Chap. 2. This technique breaks down a wide carrier into several smaller subcarriers. Data are transmitted on each of the subcarriers and then combined in to one code division channel.

5.7   Network Architecture

The fundamental difference between wireless LANs and their wired counterpart is:

- The user's address is not a fixed location. The user device can appear in any BSS and then needs to establish an association with that BSS.
- It is connected by a medium that is not physically defined.
- It is subject to outside interference.
- It is less reliable than wired media.
- It has dynamic topology.
- Unlike Ethernet it cannot be assumed that all stations can hear all other stations at any given time.
- Stations are portable or even mobile. Thus a more complex form of authentication is required.

5.7.1   Basic elements of the 802.11 standard

1. *BSS.*   The basic building block for 802.11 networks is the *Basic Service Set*. The simplest form of this is the association of two stations in an ad hoc network.

2. *DS.*   The *distribution system* is essentially the communication between a station and an access point. It handles address mapping and interworking of multiple BSSs.

3. *ESS.*   The *extended services set* allows for communication between two or more distribution services.

All of the following permutations are possible:

- The BSSs may partially overlap. This is commonly used to arrange contiguous coverage within a physical area.

- The BSSs could be physically disjointed. Logically, there is no limit to the distance between BSSs.

- The BSSs may be physically collocated. This may be done to provide redundancy.

- One (or more) IBSS or ESS networks may be physically present in the same space as one (or more) ESS networks. This may arise for a number of reasons. Two of the most common are when an ad hoc network is operating in a location that also has an ESS network, and when physically overlapping IEEE 802.11 networks have been set up by different organizations.

5.7. 2 Network components

There are two basic components to an 802.11 WLAN. The access point (AP) and the station. The access point allows for communications among stations as well as access (in some cases) to a wired LAN. The access point contains all the functions of a station (station services) as well as DSS.

Functions available in an AP are:

- Authentication
- Association
- Deauthentication
- Disassociation
- Distribution
- Integration
- Privacy
- MSDU delivery
- Reassociation

Functions available in a station device are:

- Authentication
- Deauthentication
- Privacy
- MSDU delivery

The two common types of networks are the ad hoc and the infrastructure topology.

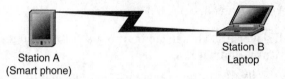

Station A
(Smart phone)

Station B
Laptop

Figure 5.1   Ad hoc network.

A WLAN formed by two 802.11 stations communicating directly with one another is called *ad hoc* (Fig. 5.1). No access point (AP) is required in this configuration. The two stations establish point-to-point connectivity.

This makes it possible for two users with compatible 802.11 stations to instantly set up a network between them. This could prove useful, for instance, in a meeting where two users wish to share data.

The infrastructure configuration involves an AP or wireless router in addition to the stations elements (see Fig. 5.2). This is a typical Small Office Home Office (SOHO) implementation. A user wishes to create a wireless network available to members of a small community. This could be family members or employees in a small business.

In addition, a number of overlapping APs can form a WLAN. This configuration not only handles portable stations but also mobile stations. That is to say a station can move from the coverage area of one AP to another (Fig. 5.3). This method allows the expansion of coverage within or around a building. It also provides additional capacity.

Lastly, an 802.11 LAN can interface with a wired LAN through the use of a router as a portal (Fig. 5.4). This allows the combination of the user's wired LAN and the Wi-Fi LAN.

5.8 Typical Wi-Fi Configurations

The following section goes over some of the issues of different Wi-Fi configurations. We started with the more simple SOHO approach and will grow to a campus environment implementation.

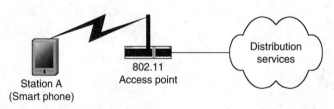

Station A
(Smart phone)

802.11
Access point

Distribution
services

Figure 5.2   Infrastructure mode.

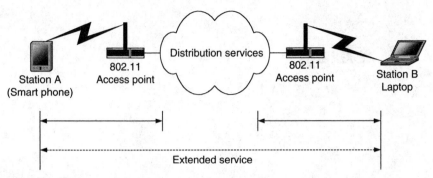

Figure 5.3 ESS.

- SOHO Wi-fi Network
- Enterprise application
- Campus deployment

5.8.1 SOHO Wi-Fi network

A sample SOHO application is depicted in Fig. 5.5. This network typically serves a home or small office, possibly only one room. There are no frequency reuse issues and probably no capacity issues. In the case of a SOHO Network there are very few design concerns.

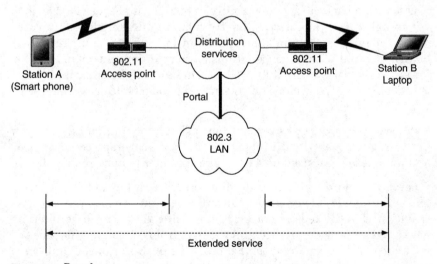

Figure 5.4 Portal.

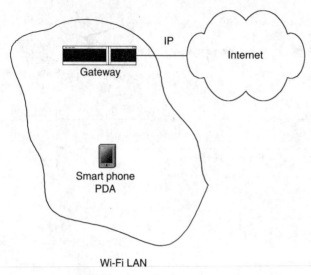

Figure 5.5   SOHO deployment.

Design Issues

Technology.   The first obvious issue is the flavor (a, b, or g) to choose. If your current equipment already has 802.11 components this may guide your decision. If not there are three basic offerings to consider. Standard 802.11b offers compatibility with many existing public networks (Hot Spots discussed later). Its compatible counterpart 802.11g offers higher data rates but is somewhat costlier.

Due to the backward compatibility of 802.11g it shares the previous advantage of Hot Spot access. Lastly, 802.11a offers higher data rates and a "cleaner" spectrum. The spectrum for 802.11a is in the UN-II band and while it is unlicensed it has more rules for control and has far fewer technologies using it at present. Due to lack of compatibility this is probably not a good choice if you intend to be mobile or portable with your Wi-Fi devices.

Coverage and capacity.   Coverage issues should be minimal unless you are trying to cover several rooms. In that case coverage could be poor and an access point used as a coverage extender may be required.

Security.   Due to the small number of users, a few simple security measures should be employed. Since the community is small, the distribution of WEP keys is not an issue. While these can be broken by hackers they provide privacy and security from casual outside users.

The use of MAC address is also practical in this situation and adds another level of difficulty to hackers. Lastly, you should customize your

SSID. This value is typically set to a default well known to hackers. While these actions are fairly simple, a vast number of Wi-Fi users don't even implement these security features.

5.8.2 Enterprise application

The task of designing a Wi-Fi network for a business or commercial application is far more complex than the SOHO example above. In addition to technical issues the enterprise will be more demanding in several areas:

- Return on investments
- Network management
- Security/monitoring
- Adaptability to capacity requirements

A number of design considerations come into play as well. An enterprise network layout is depicted in Fig. 5.6. An overview of these follows.

Design Issues

Technology.   Should 802.11a, b, or g be implemented? As before, this will be driven by a number of issues including the needs of the user community. Users should be surveyed for their requirements:

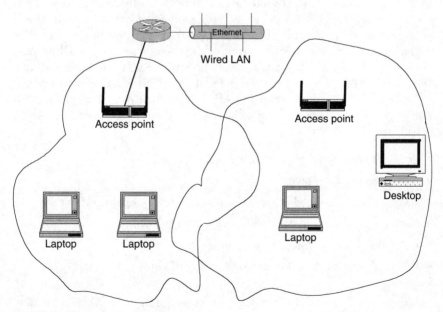

Figure 5.6   Enterprise deployment.

- Applications used
- Capacity requirements
- Current capacity used
- Mobility within the location and traveling
- Telecommuting

It may turn out that a combination of flavors will be required, with some users on 802.11g or b while other "power users" are being served by 802.11a. Power users often require higher bandwidth. By providing them essentially an overlay of 802.11a they get the BW they require and do not burden the other WLAN. If a user is a heavy traveler or telecommuter, b or g is a better choice in most cases.

Coverage and capacity.   The number and types of access points will be driven by both coverage and capacity requirements. The simple addition of access points as coverage extenders may suffice. In some areas the need for more capacity may dictate the use of multiple channels requiring rudimentary frequency planning.

Frequency planning becomes necessary when channel reuse is required. For example, in the United States there are 11 channels available to 802.11b/g. These channels overlap so that in reality only three of them can be used without interference caused by overlap. These are channels 1, 6, and 11. In the case where capacity demands multiple APs, these APs need to be set to one of these three unique channels. If more than three APs are required for capacity reuse, these channels need to maximize RF separation to minimize interference. For a more in-depth look at radio engineering issues please see Chap. 2.

In addition, a coverage survey may be required. In the simplest case the installation of an access point in conjunction with software and a laptop may be sufficient. Once the AP is up and running, signal strength measurements can be recorded throughout the service area.

Site survey issues

- Locate likely point for installing APs
- Locate available power
- Gather floor plans, electrical diagrams, and blueprints if available
- Note obstructions to coverage such as metal walls
- Establish test transmitter locations
- Install test AP and take measurements of signal strength

Security.   Some or all of the methods discussed in the SOHO example may be implemented. In the case of a large enterprise with numerous users,

WEP and MAC address techniques are too cumbersome and difficult to administer. In the case of an enterprise Wi-Fi network additional security methods need to be considered.

In addition to privacy and security the enterprise network should be monitored to detect and locate all wireless access points. It should then be determined that the APs have proper security implemented and are not simply rogue devices. In other words, make certain that they were installed and sanctioned by the IT department.

5.8.3 Campus deployment

In a campus deployment Wi-Fi access is required in two or more discrete locations. Coverage may or may not also be desirable outdoors on the campus. Much of this will be driven by the type of campus. In the case of a corporate campus outdoor coverage may be undesirable. On the other hand, a university campus may require it.

Design issues. For the most part design issues for the campus environment are very similar to those of the enterprise. There is, of course, a potentially much larger serving area in the campus environment.

The chief differentiator is the requirement for interconnecting the various locations or buildings throughout the campus. This is often accomplished through the use of additional 802.11 compliant solutions. Other methods could include wired facilities such as T1 circuits, unlicensed microwave solutions, or even optical solutions. A network diagram of a campus implementation is shown in Fig. 5.7.

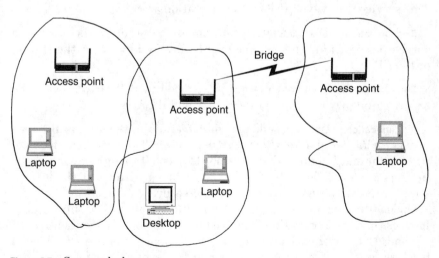

Figure 5.7 Campus deployment.

5.9 802.11 Services

The term "services" has many meanings. However, for 802.11 there are agreed upon definitions. The services defined by the 802.11 Standard address the delivery of *MAC Services Data units* (MSDU) as well as access and privacy.

The MAC services are:

Distribution services. This is the primary service used by stations to send data messages. The 802.11 Standard does not address how messages are distributed. It does, however, provide the distribution systems with enough information to deliver the message.

Integration. The integration function delivers messages from the distribution system media to an integrated LAN. This service is used when the destination or intended recipient is a member of an integrated LAN.

Association services. Association is a distribution system service that associates a station with an access point. The association service is always initiated by the mobile station. A station can only be associated with one AP (at a time), while APs can be associated with numerous stations.

Stations associate themselves with the appropriate AP based on the scanning and synchronization algorithms specified in the specification. Essentially, the station associates with an AP with the proper *Service Set Identifier* (SSID) and signal.

Reassociation. This service is used to move a station's association from one AP to another. It is used when a station is mobile. Reassociation can also be used to change association attributes while on the same AP. This service is always invoked by the station.

Disassociation. Disassociation tells the distribution service to void a current association. This service may be invoked by either the station or the AP.

Control services. Two services are introduced by 802.11 to provide the security and privacy that is inherent in wired LANs.

Authentication. Two methods are implemented in the standard—*open authentication* and *shared key authentication*. The standard does not provide for end-to-end authentication. This is left to higher levels of the network protocol. The intention is only to bring wireless to the level of wired LANs with respect to privacy and security.

Open authentication is essentially no authentication. The only requirement is that the two stations involved be set to open authentication.

Shared key authentication involves the sharing of a secret key by both stations. In 802.11 this is implemented through *Wired Equivalency*

Privacy (WEP). In this case both systems must know the WEP key in order to authenticate. 802.11 requires authentication prior to association taking place.

Preauthentication. Preauthentication can take place prior to reassociation. This speeds up the process of transitioning between APs.

Deauthentication. Deauthentication is a notification that a (non-AP) station wishes to be deauthenticated. This also results in disassociation since authentication is a prerequisite to association.

Privacy. As with authentication, the purpose of the privacy service is to bring 802.11 up to equivalency with a wired LAN. It is not intended to be a robust form of security. The privacy service is achieved using WEP mentioned earlier. If stations are set for encryption, the WEP key is used to encrypt all messages. This makes it only possible for stations sharing the key to read these messages.

Relationship between stations/services. Each station keeps track of what authentication and association level it has with all other stations. There are three levels:

State 1. Neither authenticated nor associated

State 2. Authenticated only

State 3. Both associated and authenticated

The state of a station controls what messages are acceptable. Stations will accept any message at or below the present state. In State 3 all messages are accepted; in State 2, both level 1 and 2 messages; and in State 1 only level 1 messages.

Thus only those messages allowed due to level of authentication and association will be seen by a station. If a station receives unauthorized frames it will send the appropriate deauthentication or disassociation message to the sending station.

5.10 Hot Spots

Numerous entities are developing Wi-Fi (802.11) Networks around the globe. These networks known as hot spots allow instant access to anyone with a Wi-Fi device and of course, access permission. Most of these hot spots are 802.11b systems. One group tracking the number of hot spots and their locations reports over 23,186 hot spots worldwide (jiwire.com is the source).

While many hot spots simply link the users to the Internet (provided that they have access), it is only a matter of time before more sophisticated services become available. For example, services with the ability to establish *Virtual Private Networks* (VPNs) with their inherent security between the mobile user and his or her corporate network.

It is also envisioned that roaming agreements will be extended to the wireless mobile carriers. This will enable combined cellular and Wi-Fi-capable devices to hook up to the hot spot while in its coverage area. This concept can even be extended to *handing off* to the cellular network when leaving the hot spot.

5.10.1 Access methods

There are a variety of approaches to allowing access to the hot spots:

Free access to customers: This is the current approach at some McDonalds restaurants.

Buy a burger and get a slice of time on Wi-Fi.

Monthly subscription fee: Companies such as T-mobile and Boingo are taking this approach.

Pay as you go: For example, some airline clubs give all day access for a fixed fee.

Access to current users: Verizon has installed Wi-Fi at many of the New York City phone booths; access is free to any user of Verizon's Broadband services (DSL).

A sampling of hot spot providers/operators is shown in Table 5.6. This information was taken from Internet sites in early 2004. It is of course

TABLE 5.6 Hot Spot Services

Operator	Comments
AT&T Wireless	$ 9.95/day
Boingo	$ 21.95 Monthly $ 7.95/day
T-Mobile	$ 29.99/month $9.95 per day, many Starbucks locations.
Verizon	Free to users of Verizon high-speed access
McDonalds/Toshiba	Free with purchase
STSN iBAHN	$ 9.95/day
Panera bread	Free
iPass	
Hilton hotels	
Apple stores	Free
McDonalds	In conjunction with Toshiba free trial with purchase
Surf and sip	$ 5/day

subject to constant change. The information is provided to show a variety of providers and plans. It is not an endorsement of any service.

5.10.2   Security at Hot Spots

For the most part security is nonexistent at Hot Spots. This is primarily due to the transient nature of the users and the lack of support at the locations. For example, it would be difficult to manage the use of WEP keys. It would also be somewhat useless since everyone wanting to know the key would.

5.11   Security

The intention of the 802.11 Standard is to require and specify authentication and privacy sufficient to bring wireless LANs to parity with their wired counterparts. For the most part this is accomplished through the use of *Wireless Equivalent Privacy* (WEP).

WEP provides a level of authentication that (if implemented) will prevent communications with just any station that comes along. For example, if WEP is enabled your neighbor will not be able to communicate with your LAN without the key. Of course, if he or she is a hacker it is a different story. When additional security is required other forms of security and authentication should be implemented.

In the case of mobile users VPN can be a very effective method for providing security between the mobile user and the corporate network. VPN in its many flavors is discussed later in this chapter.

5.11.1   WEP

There are two types of authentications in 802.11—*open authentication* and *shared key authentication.*

Open systems authentication.   In the case of open authentication the only requirement for authentication is that both stations have the dot11 authentication field set to open systems authentication. This is commonly used in hot spot implementation to avoid the need for distributing or sharing keys.

Shared key authentication.   Shared key authentication requires that WEP be enabled in the station. It operates using a secret shared key that has been delivered to the stations, by an alternative mechanism, to 802.11. In other words the user of the shared WEP key must obtain it prior to enabling WEP. Each user requiring access to the LAN must enter the WEP key into his or her station. Shared key authentication is invoked only if the dot11 privacy option implemented is "true." Keys are shared in a write-only *Management Information Base* (MIB) attribute.

The sequence of authentication is as follows:

- A *management message* is sent from the requestor to the responder indicating that shared key authentication has been enabled.

- A second management message is sent (from the responder to the requestor) with the128 octet generated by the WEP *pseudorandom number generator* (PRNG). This is the authentication challenge text.

- A third message is sent from the requestor back to the responder. It contains the challenge text sent by the requestor encrypted using the WEP algorithm.

- The final message denotes success or failure of authentication. The result will be "success" if (after de-encryption) the challenge text returned matches that sent in message 2. Otherwise the response will be "unsuccessful."

5.11.2 Encryption with WEP

WEP was designed to prevent casual eavesdropping. The objective was to provide privacy equivalent to that of a wired LAN. In other words if you were not on the LAN you would not see messages.

The WEP algorithm is far from unbreakable. Depending on the length of the key used, the frequency of changing the key, and the ability to key the secret WEP provides a protection from brute force attacks.

WEP is self-synchronizing and that makes it less susceptible to message delivery problems inherent in the wireless environment.

It can be implemented in hardware as well as software allowing it to be very efficient.

It is illegal to export many encryption algorithms from the United States. Every effort was made by the standards body to make WEP exportable.

Use of WEP is optional within the standard.

5.11.3 Alternatives to WEP

As mentioned earlier, WEP is not intended to provide ultimate security. In fact, it has been proven to be easily breached. It is only prudent when transmitting sensitive data to implement one or more additional forms of security.

5.11.4 SSID

Two other screening elements in the 802.11 Standard can be used to increase privacy and security. One is the SSID. Use of the SSID is intended to direct stations to the correct AP in cases where multiple APs exist. It can be used to keep casual users from accessing your network.

Unfortunately, even if the SSID is set to a secret value it can be easily learned by hackers.

In many cases users leave their SSID set to the default value. Since these defaults are widely known they leave little protection for the WLAN.

5.11.5 MAC address

This technique involves the use of the client's physical MAC address. While this is effective it too can easily be breached. The hacker simply observes network traffic and assumes the MAC address and IP address of a genuine user. Once that user drops off the network the hacker gains entry with the false ID.

In addition to the weakness of this method it is also cumbersome for an administrator to maintain.

5.11.6 Extensible authentication protocol (EAP)

With this method a client is unable to gain access to the LAN without first going through a log-in procedure the process follows:

- The client first supplies a user name and password to a RADIUS server.
- The server sends a challenge message to the client.
- The client then uses his or her password to supply a response to the RADIUS server's challenge.
- The server then authenticates the client.
- The process is then repeated in reverse.
- The client and server then create a WEP key.
- This key is delivered to the AP over the wired LAN
- The AP then sends the WEP key to the client and this key is used for the session.

Using this authentication scheme has several advantages. First of all, the use of the WEP key is reduced to only this one session. This makes WEP key discovery much less likely. Secondly, it adds an additional degree of control to LAN management.

5.12 Firewalls and Virtual Private Networks

Virtual private networks make using the Internet more like a private network. This is a concept similar to that of using WEP to make a wireless network similar, in privacy, to a wired network. Privacy and security are added to each IP packet inside another TP packet. This is known as *tunneling*. In the case of a VPN the actual data are contained inside the IP packet visible to the Internet.

A very common implementation of VPN is IP Security (IPSec). While IPSec does not provide authentication it does provide confidentiality. With IPSec, users must have an IPSec client on their PC.

Placing a VPN gateway or firewall between wireless networks and your network provides additional levels of control and security to the network.

5.12.1   Proprietary vendor solutions

There are a number of manufacturers who provide solutions to the inherent weakness in WEP. These solutions add security but usually require the use of one manufacturer's equipment throughout the network. This removes the commercial advantages of adopting an industry standard such as Wi-Fi (802.11).

5.13   Mobile VPN

What is a VPN?

The concept of Virtual Private Networks predates data networking. The term is used by traditional phone companies to refer to services offered to various entities. In some cases it is as simple as a leased line facility. In others it is a complete outsourced network that is carved out of the "public network" for an entity's exclusive use.

A virtual network is essentially a connection that appears to the users as a dedicated end-to-end connection. It is not important for the user to know the actual route taken by the data. In fact the route will often change during the course of the session. In order to make the network private (in an IP environment) other services must be implemented.

To ensure privacy several things must happen. The user must be verified to be eligible to use the VPN. This is accomplished through the use of RADIUS servers and shared certificate methods. The data must be protected or made private. This is typically accomplished through various tunneling techniques.

With a private network, users expect a certain level of service. This requires the ability to implement Quality of Service schemes. While this can be accomplished, it becomes more complex in an IP environment with varied paths and providers. In a Mobile IP environment it is further complicated by mobility and roaming issues.

5.13.1   Advantage to using VPNs

There are a number of compelling reasons to use VPN; in particular for telecommuters.

The rapid increase in telecommuters is forcing IT departments to provide remote access to more and more users. Traditionally, this access was provided through dial-up procedures; however, with ever increasing costs and the relatively low speeds, users are moving to high-speed access.

5.13.2 Increased security and control of users

Not only does VPN make corporate data private and secure, it also controls who has access to the private network. In addition, levels of access can be assigned to different users depending on their need to access sensitive information. VPN also enables the network to determine who is dialing in.

5.13.3 Cost and maintenance

Dial-up, of course, requires a modem pool. This is not only an added cost but also a source of hardware failures. Moving from dial-up to high-speed access reduces the requirements for modem pools.

5.13.4 PSTN costs

In many cases an employer supplies a form of 800 access to avoid phone charges to the employee. Obviously this cost can become quite high. For this and the reasons listed above employers are beginning to demand that telecommuters acquire high-speed access or move back to the office. The choice is clear.

5.14 Flavors of VPN

There are three basic types of VPNs: voluntary, compulsory, and chained. Each has its advantages and disadvantages. A brief description of these configurations follows. A detailed description and call flows are presented in Chap. 4.

5.14.1 Voluntary VPNs

Set up by the end user.

- Requires little intervention by the network administrators
- Hides the data from end-to-end
- Requires a public IP address
- Makes QOS implementation next to impossible

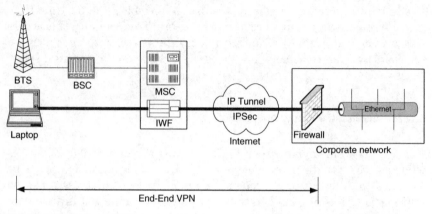

BTS   BSC   MSC   IP Tunnel / IPSec / Internet   Firewall   Ethernet

IWF

Laptop   Corporate network

End-End VPN

Figure 5.8   Voluntary VPN.

Voluntary VPN has the advantage of simplicity to the network operators. However, the need for end-to-end VPN puts special burdens on the cellular-carrier and the end-user device. The additional overhead required to take the VPN through the Radio Access Network uses scarce spectrum resources. In addition the mobile device must contain the processing power and software to implement the VPN. See Fig. 5.8.

5.14.2   Compulsory VPN

In the case of compulsory VPN any user requesting access to a network or service will use a VPN that is controlled by the operator. In the case of a cellular system this VPN may only extend from the private network being accessed to the mobile switching center. This method removes the disadvantages of spectrum utilization and end-user device requirements of voluntary VPN. It also removes the requirement for public IP addresses. Public IP addresses are becoming a very scarce resource. On the downside, a portion of the data path is unprotected as well as unpredictable. This makes the assurance of security, privacy, and level of service impossible (see Fig. 5.9).

5.14.3   Chained VPN

In the case of chained VPN essentially two unique VPN implementations are joined together. This allows, in the case of cellular, the use of a VPN method geared to the mobile environment, combined with a more prevalent VPN implementation in the nonmobile environment (see Fig. 5.10).

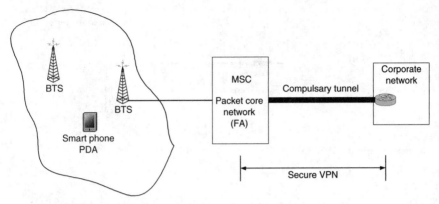

Figure 5.9   Compulsory VPN.

5.15 Mobility

5.15.1 Mobile VPN implementation in CDMA2000 networks

The core network in a CDMA2000 system is based on PPP. VPNs are typically implemented with encapsulation using methods like *Level 2 Tunneling Protocol* (L2TP). Compulsory VPNs are connected to private networks. If Mobile IP is implemented the PPP link is terminated in the CDMA operator's network at the PDSN. As mentioned earlier, if the mobile user is to be able to move from one serving PDSN to another, Mobile IP must be implemented. With Simple IP, which may suffice in many cases, the mobile will only be capable of moving within the geographic area of the PDSN it originates from.

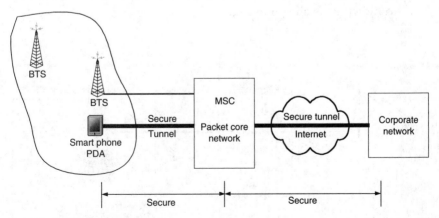

Figure 5.10   Chained VPN.

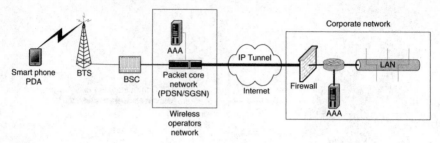

Figure 5.11 Simple IP VPN.

Due to the fact that the compulsory VPN terminates at the PDSN it is essential for the private network to understand that the security offered by the VPM extends only to the PDSN. Security from that point on is dependant on the security of a given CDMA network. While this can be assured in the case of a local user, the situation can certainly vary as the mobile roams to other carrier networks.

A brief description of VPNs in the mobile environment is included below. Details of the call set up are covered in Chap. 4. Both GSM/GPRS and CDMA2000 scenarios are discussed.

In the Simple IP case depicted in Fig. 5.11, the compulsory VPN is originated in the PDSN and a tunnel is set up between the mobile carrier PDSN and the private network LNS. In the case of simple IP the authentication and address assignment are handled by the AAA server in the private network.

In the case of Mobile IP when an MS originates a session with a private network the user must be authenticated in both the carrier's AAA server and the network's home AAA server. The tunnel established, typically L2TP, must have security provided by IPSec or some other method. This architecture is shown in Fig. 5.12.

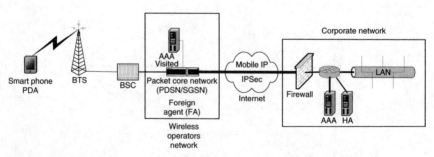

Figure 5.12 Mobile IP.

5.15.2   Voluntary VPNs in a CDMA2000 network

The option of a mobile user setting up a voluntary VPN exists but is virtually transparent to the underlying network and requires no intervention by the mobile carrier. The only requirements are those of any voluntary VPN. The MS must either have a public IP address or there needs to be an IPSec NAT transversal solution implemented.

The drawbacks to this are the additional overhead requirements on the CDMA RAN and of course the lack of true mobility of Mobile IP.

5.16   Integration with Wireless Mobile Networks

Why integrate Cellular/PCS with WLANs?

Some of the advantages of being able to transition from the cellular mobile environment to the WLAN include high bandwidth, low cost, the ubiquity of hot spots, and the offload of Cellular spectrum.

The offload of cellular spectrum may or may not also result in the reduction of cellular phone bills. With the rapid move toward almost unlimited minute talk/data plans it will be to the cellular operators benefit to move the user onto their private WLANs when possible.

Wi-Fi and other WLANs offer high-speed data transfer within very specific and usually small areas. Wireless mobile systems, on the other hand, provide a far wider area of service (global). The data rates with wireless mobile are however significantly slower. This makes the integration of the two access technologies highly complimentary.

The sticking point at present is, "Who and how will Wi-Fi operators—both corporations and hot spots—team up and how will they combine services with mobile operators? How will they bill for the service? Will subscribers pay for data access on the unlicensed spectrum?"

The following section details two potential scenarios for combining Wi-Fi and mobile service. The first example is defined as "loosely coupled." In this implementation the user experiences what appears to be ubiquitous coverage from the mobile carrier. However, the actual data session (on the Wi-Fi Network) is independent of the mobile operator's core network. There is, of course, the need for authentication of the user. The basic configuration for "loosely coupled" integration is shown in Fig. 5.13.

The tightly coupled case has the Wi-Fi LAN directly connected to the mobile operator's core packet network. It functions as just another 3G network in the eyes of the mobile network. Some of the drawbacks include the difficulty for independent hot spots and corporate networks to be part of this implementation due to the direct access into the 3G core network. As shown if Fig. 5.14, AAA services are combined in this implementation.

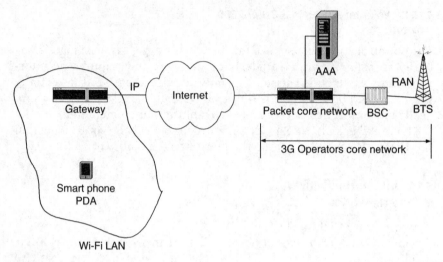

Figure 5.13 Loose integration.

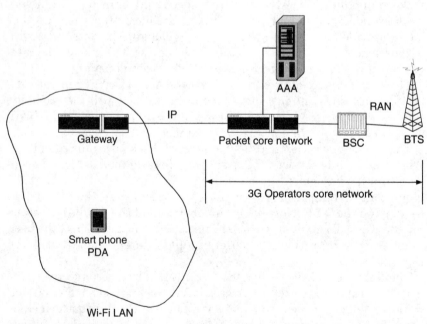

Figure 5.14 Tight integration.

5.17 Conclusion

An overview of Wi-Fi (802.11) Standards was presented in this chapter. This set of standards has made it possible to develop wireless LAN solutions that are transparent above the MAC, PHY, and link layer; enabling the proliferation of multivendor 802.11 solutions, in particular the 802.11a,b, and g family. This not only makes solutions widely available, they are also economical and interoperable.

Due to these facts Wi-Fi networks are becoming available in homes, small offices, private corporations and many public Hot Spot networks. This widespread availability is driving a de facto acceptance of Wi-Fi for WLAN. The devices are becoming almost standard in laptops and will eventually be as common as NIC cards once an option on most PCs.

This widespread adoption and availability make the development of mobility mandatory.

This mobility is enabled to a large extent by mobile VPN. Mobile VPN enables the seamless operation of mobile 3G wireless networks and, as we have seen, can enable the integration of Wi-Fi and 3G networks.

One of the last unknowns is how will these services be linked together commercially. Subscribers have not typically been that attracted to a bundled bill for services unless there is an associated saving. The idea of bundling 3G mobile services with Wi-Fi, particularly hot spots, seems like an obvious migration. The big question is who will own the customer?

References

http://www.IEEE.org/

802,®-2001, "Revision of Institute of Electrical and Electronics Engineers Std 802-1990," Institute of Electrical and Electronics Engineers.

802.11, [ISO/IEC 8802-11: 1999], American National Standards Institute/Institute of Electrical and Electronics Engineers, 1999.

802.1B and 802.1k [ISO/IEC 15802-2], "LAN/MAN Management,"American National Standards Institute/Institute of Electrical and Electronics Engineers.

802.1D [ISO/IEC15802-3], "Media Access Control (MAC) Bridges,"American National Standards Institute/Institute of Electrical and Electronics Engineers.

802.1F, "Common Definitions and Procedures for Institute of Electrical and Electronics Engineers 802 Management Information," Institute of Electrical and Electronics Engineers.

802.1G [ISO/IEC 15802-5], "Remote Media Access Control (MAC) Bridging,"American National Standards Institute/Institute of Electrical and Electronics Engineers.

802.2 [ISO/IEC 8802-2], "Logical Link Control,"American National Standards Institute/Institute of Electrical and Electronics Engineers.

Smith, C. and D. Collins, *3G Wireless Network*, McGraw-Hill, New York, 2002.

Muller, N. *Wi-Fi for the Enterprise*, McGraw-Hill, New York, 2002.

Minoli, D. *Hotspot Networks*, McGraw-Hill, New York, 2002.

Ohrtman, F. and K. Roeder, *Wi-Fi Handbook*, McGraw-Hill, New York, 2003.

Miller, S. *Wi-Fi Security*, McGraw-Hill, New York, 2003.

Shneyderman, A. and A. Casati, *Mobile VPN*, Wiley, New York, 2003.

IEEE 802.16 (WMAN/WiMax)

IEEE Standard 802.16, or 802.16, is a wireless protocol that focuses on the last-mile applications of wireless technology for broadband access. 802.16 is referred to as Wireless MAN and a subcomponent of the standard is called WiMax, which falls under 802.16a. Therefore 802.16 is a set of evolving IEEE standards that are applicable to a vast array of the spectrum ranging from 2 GHz to 66 GHz, which presently include both licensed and unlicensed (license exempt) bands.

Fundamentally, 802.16 is the enabling technology or standard that is intended to provide *Wireless Metropolitan Area Network* access to locations, usually buildings, by the use of exterior illumination typically from a centralized base station as shown in Fig. 6.1. The simplified system shown in Fig. 6.1 could also have multiple subscribers referred to as *subscriber stations* (SS) and connected back to the base station via the 802.16x RAN.

802.16 is a Point-to-Multipoint Protocol, that is a standard providing equipment manufacturers and operators with a standard for access profiles as well as known interoperability levels allowing for multivendor environments.

The point-to-multipoint is a concept where multiple subscribers can access the same radio platform using both a multiplexing method as well as queuing. 802.16 systems operate in the microwave frequency band and use similar radio technology as a Point-to-Point Microwave System.

The Point-to-Multipoint Protocol used is a connection-oriented system that can take on a star or mesh configuration using both FDD and TDD. In addition, 802.16 is in itself protocol independent in that it can transport both ATM and IP depending on the content desired to be ported. It also uses both contention and contentionless access supporting services that are AAL1 to AAL5.

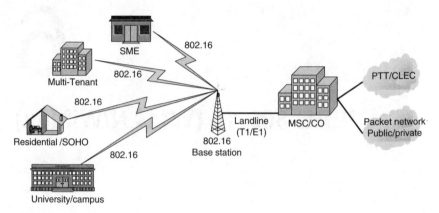

Figure 6.1   802.16 general configuration.

A Point-to-Multipoint System enables an operator to handle more subscribers or rather Mbps/km^2 than a microwave Point-to-Point System using the same amount of radio frequency spectrum.

The system configuration used for an 802.16 system is designed to operate efficiently within the spectrum that is allocated. To achieve this, 802.16 is a wireless system that employs cellularlike design and reuse with the exception that there is no handoff. It can be argued that 802.16 is effectively another variant to the LMDS and WLL portfolio described previously and referred to as *proprietary radio systems*.

Local Multipoint Distribution Systems (LMDS), *Fixed Wireless Point-to-Multipoint* (FWPMP), and *Multichannel Multiple Point Distribution Systems* (MMDS) have been in existence for quite some time now. However LMDS, FWPMP, and MMDS—while being superb in delivering a vast array of broadband services—have suffered from the proprietary systems that have not seen the reduction in the total cost of deployment or ownership being reduced to a level that competes with the existing wired broadband services.

Wireless Internet Service Providers (WISPs) have also been active in delivering broadband services. WISPs have used both licensed and license-exempt bands for service delivery. WISPs have been using Point-to-Multipoint or mesh systems along with 802.11 to serve their customers.

However, 802.16 is different from 802.11 and wireless mobility systems like GSM, CDMA, and UMTS. 802.16 is a unique wireless access system whose purpose is to provide broadband access to multiple subscribers or locations within the same geographic area. It uses microwave radio as the fundamental transport medium and is not fundamentally a new technology but rather an adaptation and standardization of existing technology for broadband service implementation.

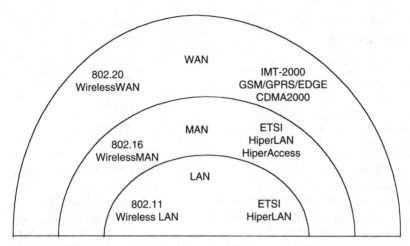

Figure 6.2 802.16 position in network deployment.

For a wireless service provider 802.16 is one of the many tools that can be used to help improve the network and drive operating and capital costs downward. 802.16 is one of the many 802 specifications that have so profoundly influenced society.

Figure 6.2 is an illustration that shows the relative position or place 802.16 has within several of the 802 standards, specifically 802.11 and 802.20. It is important to note that 802.16 addresses the transport layer for the MAN and enables end devices to be aggregated into a larger pipe for overbooking, leading to cost reductions.

802.16 must accommodate both continuous and bursty traffic and therefore is designed as a set of interfaces that are predicated on a common *Media Access Control* (MAC) Protocol. The MAC design also factors in it several physical layers that are needed due to spectrum availability, utilization, and regulatory requirements.

802.16 systems can be effectively deployed where:

- Users are dissatisfied with the current packet and/or network interface
- Network operators need to reach customers cost effectively
- New service offerings are needed for 2.5G/3G augmentation

To achieve this, standards are the key to enabling 802.16 in becoming a cost-effective backhaul and transport layer for a wireless mobility system.

Some of the more salient advantages of a wireless system for broadband access to use the 802.16 Standard are:

- Bandwidth on demand (BOD)
- Higher throughput
- Scalable system capacity
- Coverage
- Quality (CBR and UBR)
- Cost (investment risk and end user fee)

802.16 can and will provide a viable method ensuring the connectivity back from access points (802.11) to the service provider, i.e., the last-mile pipe.

In short, 802.16 will enable operators to pick from multiple vendors for their network architecture. Therefore the driving issue for wireless providers to deploy any system using 802.16 is to provide a more cost-effective access method either to connect the cell sites together or provide backhaul for Wi-Fi hot spots. The 802.16 model, for wider acceptance, follows in the footsteps of the proliferation of the 802.11b and g *customer premise equipment* (CPE) from both a network interface and an access point and has been a leading contributor to the success of this access technology.

6.1 802.16 Standard

The IEEE 802.16 Standard is a wireless *Metropolitan Area Network* (MAN) technology that provides both backhaul and an alternative for last-mile broadband access as well as connecting 802.11 hot spots to the Internet. There are several standards presently that are within the general 802.16 specification as of this writing with promises of more to come. The technology platforms that adhere to the 802.16 Standard will provide broadband wireless connectivity to fixed, portable, and nomadic devices.

802.16-compliant equipment are meant to operate in multiple spectrum plans around the world. The two major spectrum channel plans are defined by the *Federal Communications Commission* (FCC) and *European Telecommunications Standards Institute* (ETSI). The channel plans, in addition to spectrum locations, are different for both the FCC and ETSI allocations. The typical FCC allocation for broadband systems is 20 or 25 MHz while for Europe the spectrum allotted is 28 MHz with both allocation methods favoring duplexed operation. The standard also supports *Frequency Division Duplexing* (FDD) and *Time Division Duplexing* (TDD). The inclusion of TDD is aimed at regulatory environments where structured channel pairs do not exist.

The FCC channel allocation plan is done in increments of 5 MHz ranging from 5,10,15, 20, and 25 MHz channels that can be either FDD or TDD in nature. The FCC channel plan and spectrum also need to adhere to 47 CFR Part 101 as well as parts 1, 15, and 17.

The ETSI channel plan for 802.16 has different channel bandwidths or channel plans than that used in the United States. The channel plans are based on increments of 3.5 MHz ranging from 3.5,7,14, 28, 56, and 112 MHz, depending on the country and spectrum that the system is to operate within.

The 802.16 Standard has many fundamental properties

1. It supports multiple services simultaneously

2. Bandwidth on demand

3. Link adaptation (4QAM/16QAM/64QAM)

4. Point-to-point topology integrated with mesh topology

Table 6.1 illustrates the different bands that the different 802.16 Specifications occupy.

802.16 has strived to ensure compatibility with both the FCC and ETSI allocations. With that in mind, 802.16a—also known as WiMAX— has the same access method as HiperMAN and promises increased roaming and flexibility in fixed wireless.

In general, 802.16 is a Point-to-Multipoint Protocol with a centralized base station and all the 802.16 Standards draw upon the DOCSIS Protocol for defining its service scheduling techniques.

- 802.16 Standard specifies two convergent service layers that form the basis of protocol. The two convergent service layers are ATM and Packet (IP).

- 802.16a focuses on the spectrum that is below 10 GHz. In the United States the key spectrum for 802.16 is in the MMDS bands, mostly from 2.5 to 2.7 GHz. Worldwide, 802.16a is meant for the 3.5 GHz and 10.5 GHz bands since they are seen as good prospects for residential and small-business services.

- The 802.16e Standard's purpose is to add limited mobility to the current standard which is designed for fixed operation. IEEE 802.16e is not

TABLE 6.1 802.16 Bands

Standard	Band	Comments
802.16	10–66 GHz	WirelessWAN, HiperAccess
802.16a	2–11 GHz	WiMAX, HiperMAN Licensed bands
802.16a (formerly b)	5–6 GHz	Unlicensed band (Mesh)
802.16e		Nomad

intended to compete with 3G or other truly mobile efforts. Line of sight (LOS) is required for 802.16 but becomes less of a factor below 10 GHz due to the use of adaptive antennae and modulation format variations.

The uplink and downlink paths both support and use adaptive bandwidths or link adaptation allowing for better traffic and spectrum management.

There are two types of customer premise equipment classifications for 802.16. The two types are defined as *grant per subscriber station* (GPSS) and *grant per connection* (GPC). Each is defined based on its ability to accept or use the available bandwidth that is allocated or available to it. The GPSS enables the aggregation of services by SS while the GPC does it on a per connection method.

There has been extensive work done in the standards regarding the protection of information and ensuring that access is not compromised. The authentication and registration of the 802.16 SS is done by X.509 digital certificate which is provided at the point of manufacture. Additionally, encryption is also performed on the links to ensure data security through *Data Encryption Standards* (DES).

For a more detailed read on the specifications related to 802.16 there are several excellent sources for obtaining information and existing vendor interoperability.

- www.ieee802.org
- www.wimaxforum.org

6.2 Design Considerations

The design process used for any 802.16 system involves many aspects that draw upon both microwave point-to-point system and Cellular/PCS system designs. However, there are unique aspects to 802.16 systems, which while drawing upon other system designs that are well published use a combined or rather hybrid approach.

It is important to note that the initial 802.16 system design process is extremely fluid and is equally complex when addressing system-growth issues. The design process can be extremely simplistic or elaborate as time and management allow. Since 802.16 involves selectively covering areas where the customers are, the benefit that they are effectively stationary leads to the ability to significantly reduce the complexity in site location and design.

However, just putting up a single boomer site while satisfying the initial system requirements for possibly covering the initial customer or customers may, in fact, limit the overall throughput or capacity carrying ability of the system.

Included with the design process is the access terminal location. Primarily, the system design needs to be done such that the alterations to access terminal locations, where the customers are actually connected to the system be minimized. While the access terminal alteration issue cannot be completely resolved in every case, care must be taken in the initial design to ensure that this issue is minimized in the best practically possible way.

It can never be overemphasized in the initial design phase that the amount of traffic expected to be handled by any one site or even the system in total be reviewed prior to implementation. The design review stage is the best point in the design process to screen potential site problems. The initial capital authorization for the system itself should verify that the capacity envisioned for the system using the expected services offered with the appropriate tariffs applied should prove the economic viability of the system. For example, if a site is only needed for handling a total of 256 kbps of data and voice traffic combined, the solution for that geographic location may be reselling landline service or abandonment. A similar analysis can be carried forward at the system level.

With that said, the primary question that any design engineer is faced with when beginning the design is determining where to begin. In order to begin the design, the fundamental concept of what the purpose or criteria for the design is needs to be established. Then the issue of what will be the output of the project and when this information will be released, and in how many steps needs to be taken up.

There are numerous design considerations that need to be factored into the possible use and deployment of an 802.16 system. The decisions relate to the economical deployment and operation of the system.

However, the fundamental issues that need to be addressed for the deployment of an 802.16 wireless system or any wireless system are

- What are you trying to accomplish (i.e., the objective derived from the business plan)?
- Why are you doing it?
- When does it need to be operational (i.e., the schedule)?
- How is this process going to be done?

Answering the four basic items above will go a long way in helping define and deliver a system.

There are several components to 802.16 and depending on the application one or several protocols can be used, along with the associated network equipment, to deliver the requisite service.

Table 6.2 is a brief overview of the protocol to be used for the application that the design is being applied to.

TABLE 6.2 802.16 Specifications

Application	Specification (solution)
Fixed	802.16
Fixed and portable	802.16a
Nomadic	802.16e

To further help in the design selection process Table 6.3 highlights some of the more salient issues between the various 802.16 Standards. It is important to note that 802.16 and 802.16e focus primarily on licensed spectrum while 802.16a includes components that include both licensed as well as unlicensed spectrum.

6.3 Topology

The overall topology of the wireless network using 802.16 depends on several factors, some of which are depicted by the spectrum allotted, capacity needed, services delivered, and of course the physical implementation issues. The two primary topologies used are either star or mesh. The star configuration is shown in Fig. 6.3 and illustrates a potential backhaul configuration for a wireless mobility system using 2.5G or 3G technology.

The mesh configuration shown in Fig. 6.4 would be more often associated with a wireless mobility system supporting 802.11 wireless access points—Wi-Fi hot spots. Each configuration has its advantages and disadvantages. The star configuration would support more capacity, Mbps/km^2, than the mesh configuration. However, the mesh configuration would enable a more rapid deployment and not rely on LOS with the main hub cell where traffic concentration takes place.

TABLE 6.3   802.16 Standards Comparison

	802.16	802.16a	802.16e
Spectrum	10–66 GHz	2–11 GHz	2–6 GHz
Channel bandwidths	20, 25, and 28 MHz	1.5 to 20 MHz	1.5 to 20 MHz with UL sub channels
Modulation	QPSK/, 16QAM, 64QAM	OFDM 256 sub carriers QPSK/, 16QAM, 64QAM	OFDM 256 sub carriers QPSK/, 16QAM, 64QAM
Bit rate	32–134 Mbps (28 MHz channel)	75 Mbps (20 MHz channel)	15 Mbps (5 MHz channel)
Channel conditions	LOS	Non-LOS	Non-LOS
Typical cell radius	2–5 km	7–10 km, max 50 km	2–5 km

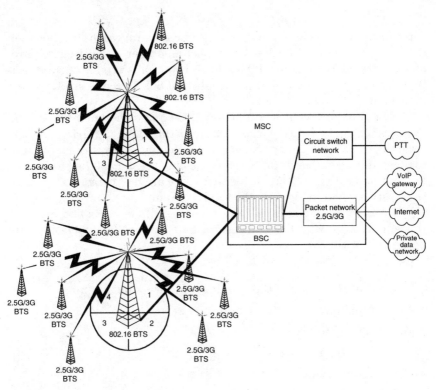

Figure 6.3   802.16 star configuration.

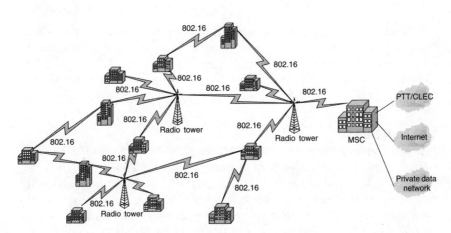

Figure 6.4   Mesh configuration.

6.4 Coverage

Any wireless technology requires a good communication link between the sending and receiving devices to function properly. Coverage requirements vary depending on which 802.16 Protocol is used in the application. A quick glance at Table 6.3 reveals that 802.16 requires LOS while 802.16a and 802.16e do not require LOS.

The LOS requirement is a function of the radio spectrum the system operates within. More to the point 802.16 operates in the 10 GHz to 66 GHz band that requires LOS for proper communication. Many tests have been conducted in these bands and LOS is the only reliable method for ensuring communication. LOS requires that there are no Fresnel Zone infractions between the sender and the receiving units.

When LOS is not required, as is the case with 802.16a and 802.16e, this simply means that an indirect path or Fresnel Zone violations are permitted, within reason.

Coverage for the site is dictated by

- Height above ground of the sending and receiving antennae
- Terrain (mountains, valleys, and the like)
- Morphology (buildings, trees, and the like)
- Antenna gain and ERP
- Modulation format
- Link reliability (ITU rain zone)
- Polarization
- Throughput requirements

Ultimately the question comes about to ask what the range of the base station will be. The answer for a point-to-point link or 802.16 is based on path loss that is dictated by free-space loss. The 802.16a and 802.16e coverage, however, is more a function of propagation path loss that is in the order of $1/r^4$ or $40 \log_{10}$.

6.5 Link Budget

The link budget used for an 802.16 System design is one of the key technical parameters used for the design process. Though all the components in the design process are important, the link budget directly determines the range and deployment pattern used for a given system. An important aspect is that there can be several link budgets in any system based on the operating frequency, spectrum allocated, link reliability, physical components for the system, differences in up and down stream modulation methods or protocols, and rain fade issues, to mention some of the elements that need to be considered.

The objective behind the link budget is to determine the path length or rather the size of the cell sites needed for the network design.

As with all wireless technologies, the available throughput is a function of several key elements.

- Radio Access Protocol
- Bandwidth available
- Radio Link Quality
- Backhaul

The Radio Access Protocol is a determining factor in that you should apply the correct protocol to the application at hand because no single protocol is completely universal. Bandwidth available will determine the capacity and hence the services, both individual and aggregated, that can be supported by the radio link.

An example of a link budget is included in Chap. 2.

The radio link quality will directly alter the throughput available. As with all wireless access systems, the lower the radio link quality the lower the throughput allowed or possible. The radio link quality is a function of coverage and interference coupled with modulation format used in an allotted bandwidth and spectrum. Figure 6.5 is an attempt

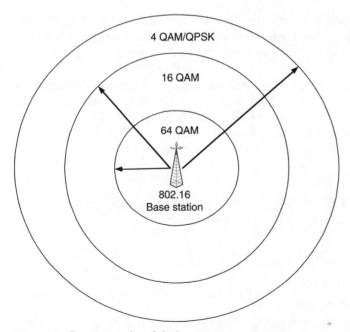

Figure 6.5   Coverage and modulation.

TABLE 6.4   Throughput Chart

	QPSK	16QAM	64QAM
Channel bandwidth MHz	Bit rate Mbps	Bit rate Mbps	Bit rate Mbps
20	32	64	96
25	40	80	120
28	44.8	89.6	134.4

to depict conceptually the relationship between modulation scheme and the radius or rather coverage of an 802.16 site. Figure 6.5 illustrates that as the site tries to cover more geographic area, a tradeoff is made in terms of throughput allowed due to the lower modulation scheme employed.

To further drive home the relationship between coverage and modulation scheme, Table 6.4 shows a few examples of maximum throughput, as a function of both modulation scheme as well as discrete channel bandwidths for 802.16. The modulation format used is also defined to be dynamic and adaptive, to account for varying QoS and network bandwidth as well as link quality variations.

6.6   Network Considerations

There are multiple network configurations possible with an 802.16 wireless access system. The configuration that the network takes on is dependant on the services offered and/or transported—resiliency, growth, and flexibility.

Some of the services that an 802.16 system can support are listed for quick reference. It should be noted that the services listed are not all inclusive of what can be delivered or will be delivered. The only exception is that the service offered cannot have a bandwidth requirement greater than what the radio transport layer can support.

- LAN/WAN (VPN)
- T1/E1 replacement (clear and channelized)
- Fraction T1/E1 (clear and channelized)
- Frame relay
- Voice telephony (POTS and enhanced services)
- Video conferencing
- Internet connectivity
- Web services (email, hosting, virtual ISP, and the like)

- E-commerce
- VoIP
- FaxIP
- Long distance and international telephony
- ISDN (BRI and PRI)

For 802.16a and 802.16e the services center around IP and do not include Constant Bit Rate (CBR)-type traffic as 802.16 does. Regardless, the host of services and perturbations to those just listed above, make an impressive portfolio to offer. Of course the necessary platforms and connectivity for the network need to be in place in order to ensure that these services can and will be able to be offered and effectively delivered.

802.16 systems, as like all other networks, can have both on-net and off-net traffic considerations. Ideally the traffic should be all on-net but when the system initially goes online most, if not all, of the traffic goes off-net and the use of the PTT or another CLEC will be required to facilitate the delivery of the service almost exclusively.

6.7 Network Configurations

There are several general configurations for 802.16 wireless access systems as they apply to wireless mobile systems. The wireless mobile systems that can take advantage of the robustness of 802.16-defined systems fall into the 2.5G and 3G Network platforms that use a packet network for delivering services.

The following are a few of the many possible network configurations that are possible with the use of 802.16-specified equipment.

- Cellular backhaul
- Last-mile broadband (802.11 in replacing DSL/cable)
- Hot spot bandwidth provider
- Residential service (WLL)
- Unservered areas (Greenfield)

6.7.1 Cellular backhaul (802.16)

The use of 802.16 for T1/E1 replacement or augmentation can take on the configuration shown in Fig. 6.3. The configuration in Fig. 6.3 is a star configuration having multiple BTSs connected to a few 802.16 hub sites that act as traffic concentrators.

Historically, 802.16 by itself for wireless backhaul needs to have the economics of the transport method reviewed in detail. However, when T1/E1 facilities are not available or the cost structure of those circuits has not been lowered due to competition, the possibility of using 802.16 for backhaul becomes more compelling.

In addition, the use of 802.16 can be made to augment the resiliency of the network for critical paths or sites. Typically, a radio site is connected to a single central office and even though it may be on a fiber ring the last 1000 ft is usually a single point of failure. Therefore 802.16 can be used to provide not only backhaul but also alternative routes for T1/E1 facilities.

6.7.2 Wireless packet network offload (WiMAX/802.16a)

WiMAX access points, if deployed properly, can help directly offload the wireless mobile system by backhauling packet data over the 802.16a RAN instead of the 2.5G/3G RAN.

Figs. 6.6 through 6.7 are used to illustrate an application where 802.16-specified equipment can help offload the existing mobile wireless system from high data usage requirements and thereby free up valuable resources.

The figures show an 802.16 link for backhaul connecting the BTS to the BSC. However, depending on the facilities available and their cost, the use of an 802.16 system for backhaul may not be necessary.

The situation is where a subscriber using one of the various 2.5G/3G services begins or wants to begin using bandwidth-intensive services. The bandwidth-intensive services can cause a drain on the available resources of the wireless mobility system. Alternatively, the customer wants to place a voice call using either the wireless provider or their corporate iPBX, and migrating them to the 802.16a system would also free up valuable resources.

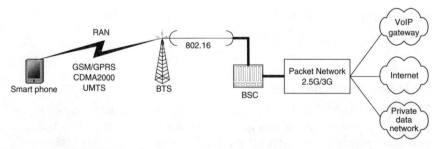

Figure 6.6   2.5G/3G Packet data.

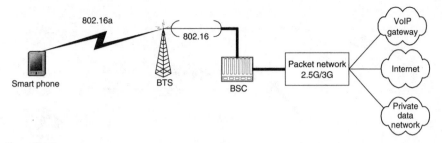

Figure 6.7   802.16a 3G packet data.

Therefore in Fig. 6.6 the subscriber begins the data session request. The system then hands off the subscriber to the 802.16a system using mobility management from the 2.5G/3G packet system to the 802.16a network in Fig. 6.7. Alternatively, if the area was a Wi-Fi hot spot then the transition would be to an 802.11 access node that could be connected back to the wireless mobile network using 802.16a as shown in Fig. 6.8.

6.7.3 High-speed wireless packet (802.16e)

The introduction of 802.16e equipment into the network can be used for campus environments where the mobility to the 2.5G/3G network is not needed. Instead, the service offered is the ability to provide Packet Data Network Services and have traditional voice services offered via legacy devices.

Figure 6.9 illustrates a subscriber accessing the 2.5G/3G packet data network via an 802.16e link. The unique feature of the 802.16e protocol is to not only provide broadband but also provide for mobility within the campus via handoffs to ensure that the session is not interrupted. An example of a handoff is shown in Fig. 6.10 where the subscriber moves from BTS(a) to BTS(b) while maintaining the packet session.

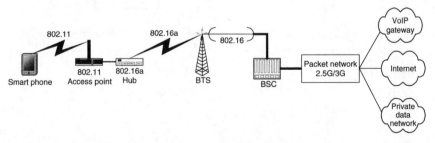

Figure 6.8   802.16 and 802.11.

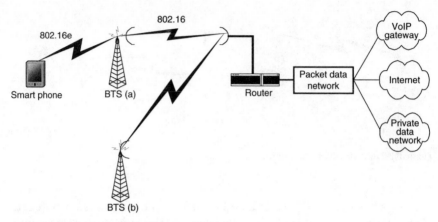

Figure 6.9   802.16e simple campus configuration.

6.8   Integration with 2/5G/3G and 802.11 Mobility

An obvious question would begin to arise regarding the integration of 802.11 and 802.16 and that is, "why not use 802.11 instead and keep one type of infrastructure?" The introduction of 802.16/16a and 16e into a mobile wireless network can run counterintuitive to the simplification process.

802.16 can help reduce churn by enticing enterprise customers and fleets with higher bandwidth, VoIP integration, and remote access services. In addition, the advantage that wireless mobile operators have by

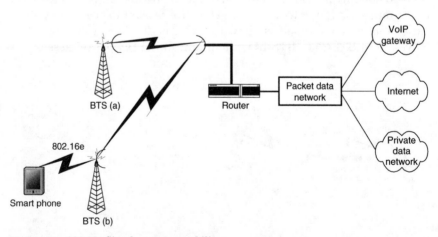

Figure 6.10   802.16e Simple campus mobility.

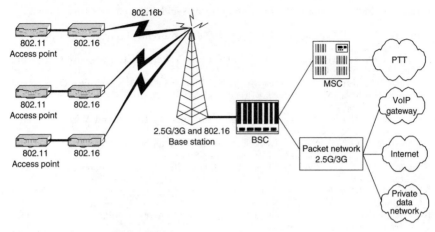

Figure 6.11   Access point backhaul.

using 802.16/a/e in their systems would be the freeing up of 2.5G/3G radio spectrum for use by other customers who are truly mobile and do not require the use of packet data services.

802.16 has the advantage of providing a more cost-effective link for connecting several BTSs or more importantly several 802.11 access nodes back to the core packet network. In short, 802.16 advantages over an 802.11 system involve:

- Multiple QoS, not just contention based
- Higher data rates
- Longer distances (access node range)

The rationale is that 802.16a was envisioned for WMAN while 802.11 was designed for local wireless LAN requiring connectivity to be effective.

However, the biggest factor for integrating 802.16 and specifically 802.16a into a 2.5/3G Network is the ability to transport the IP traffic generated on the 802.11 Wi-Fi Networks as shown in Fig. 6.11.

References

802.16-2001, "Part 16: Air Interface for Fixed Broadband Wireless Access Systems," IEEE Standard for Local and Metropolitan Area Networks, April 8, 2002.

802.16t-00/02, "Coded Orthogonal Frequency Division Multiple Access," Runcom Technologies Inc, 2000.

Beyer, D., and C. Eklund, S 802.16a-02/30, "Tutorial: 802. 16 MAC Layer Mesh Extensions Overview," 2002.

Chang, D., and S. Varma, "WiMax Rolls Interop Guidelines for 802.16a," *Electronic Engineering Times*, May 16, 2003.

Clinkert, A., IEEE S802.16e-03/23, "802.16e Requirements from an Operator's Perspective," 2003.

Cornelius, J. C., IEEE 802.16hp-00/04, "FCC Rules in Regards to Unlicensed Spectrum Usage from 2 to 11 GHz," 2000.

Eklund, C., R. Marks, and K. Stanwood,"IEEE Standard 802.16: A Technical Overview of the Wireless MAN Air Interface for Broadband Wireless Access," *IEEE Communications Magazine*, June 2002, pp. 98–107.

Gallagher, V., IEEE 802.16hc-00/02, "Requirements for Broadband Wireless Access Systems in the UNII Bands," 2000.

Kasslin, M., and N. V. Waes, IEEE 802.16hc-00/09, "Applicability of IEEE802.11 and HIPERLAN/2 for WirelessHUMAN Systems," 2000.

Kasslin, M., and N. V. Waes, IEEE 802.16hc-00/01, "Requirements for WirelessHUMAN Systems," 2000.

Khanna, C., IEEE 802.16hc-00/10/, "Comparison of Existing and Proposed Wireless Standards," 2000.

Kostas, D., IEEE 802.16 hc-00 /05, "USA Regulatory Constraints of a U-NII Broadband Wireless Access System Standard," 2000.

Kostas, D., IEEE 802.16hc-00/07, "Specifying TDD for the Proposed WirelessHUMAN Standard", 2000.

O'Keefe, S., "TDD vs. FDD: The Next Hurdle," *Telecommunications*, December 1999, p. 40.

Peha, J. M., IEEE 802.16hc-00/03, "The Path Towards Efficient Coexistence in Unlicensed Spectrum", 2000.

Segal, Y., and Z. Hadad, IEEE 802.16hc-00/08, "Contribution for the WirelessHUMAN PAR," 2000.

Segal, Y., Z. Hadad, I. Kitroser, and Y.Lieba, IEEE 802.16.4c-01/35, "OFDMA for Mesh Topology," 2001.

Segal, Y., Z. Hadad, I. Kitroser, and Y. Lieba, IEEE 802.16.4c-01/34, "OFDMA Advantages for the 802.16b," 2003.

Sellars, M., and D. Kostas, IEEE 802.16hc-00/06, "Comparison of QPSK/QAM,OFDM, and Spread Spectrum for 5-6 GHz PMP BWAS," 2000.

Smith, C., *LMDS*, McGraw-Hill, New York, 2000.

Sydor, J., IEEE 802.16hc-00/12, "A Proposed High Data Rate 5.2/5.8-Ghz Point-to-Multipoint MAN System," 2000.

Wirbel, L., "LMDS,MMDS Race for Low-Cost Implementation," *Electronic Engineering Times*, November 29, 1999, p. 87.

www.ieee802.org

www.wimaxforum.org

www.wirelessman.org

802.20—MBWA

The IEEE 802.20 specification, *Mobile Broadband Wireless Access* (MBWA) or 802.20 for short, can be considered to be the next evolution in the quest to provide one platform for mobile wireless services. 802.20 is a broadband mobile access technology which will compliment the IMT2000, 3G, and 2.5G platforms that exist throughout the world today.

802.20 is different from 3G and 2.5G platforms in that the protocol used is optimized for IP services and does not have the legacy issues to support circuit-switched services like 2.5G and 3G. The 802.20 specification is universal and works within the existing IMT2000 mobile wireless bands throughout the world. The fundamental goal of the specification is to enable this access platform to be deployed worldwide providing interoperability for broadband wireless access that will also be based on a true open standard interface, unlike the current proprietary platforms.

802.20-compliant equipment is designed to operate in the licensed wireless mobility spectrum that is defined in the IMT-2000 specification. The specification is designed to be optimized for IP-data transport enabling peak data rates of greater than 1 Mbps for mobility (not stationary) and will support vehicular speeds of up to 250 km/h, meaning it will interface with mobile satellite systems.

The 802.20 specification is also different from the 802.16 series of specifications in that it is meant for mobility and not a modification of fixed wireless access that is modified for mobility. The specifications and details of the protocol can be found at www.ieee.org.

7.1 Migration Path

While being a new specification, 802.20, was built from the ground up with coexistence with the existing 2.5G/3G Networks being a must. The 802.20-compliant equipment is meant to share the existing 2.5G/3G

TABLE 7.1 Wireless Technology Platforms

Wireless generation	Access technology
1G	AMPS, NAMPS, TACS/ETACS
2G	IS-136, IS-95, iDEN, GSM, DECT, LMR
2.5G	CDMA-2000 (1xRTT), GSM/GPRS, GSM/EDGE, WiDEN
3G	CDMA-2000 (EV-DV), W-CDMA (UMTS), SCDMA

infrastructure, where possible. The systems will be integrated as an overlay onto the wireless mobility network.

802.20 systems are designed for IP and mobility without the legacy platform constraints, offering a more ubiquitous service and lower cost of deployment and ownership for Mobile IP, for both Simple and Mobile IP including VPN.

The legacy wireless mobility systems that will be able to integrate into or coexist with an 802.20 wireless system include the following overall systems indicated in Table 7.1.

Figure 7.1 best represents the overall flow or rather interaction between the various wireless standards for a migration path to 802.20. The general flow follows the IMT-2000 migration path format with the convergence of wireless mobility taking place at 802.20.

7.2 Technical Parameters

The technical parameters or specifications for 802.20 are still in the making at this time. However, 802.20 is specifically designed to integrate into the existing IMT-2000 series of mobility platforms shown in Fig. 7.2.

The 802.20 modulation scheme will most likely be OFDM for the purpose of being more spectral efficient as well as integrating into the 1.25 MHz or 5 MHz TDD or FDD bands that are IMT-2000 compatible. The MAC structure is still in development. However, the following specifications in Table 7.2 will be associated with 802.20. Additionally, the data rates shown in Table 7.2 represent estimated downlink and uplink rates using 2.5 MHz of radio aggregate spectrum, whether consumed via an FDD or a TDD system.

Because the 802.20 frequency allocation scheme includes both paired and unpaired channel plans with multiple bandwidths, e.g., 1.25 or 5 MHz, it enables deployment within existing mobile wireless systems. The 802.20 channel bandwidths are consistent with frequency plans and frequency allocations enabling a wide area deployment into the existing spectrum.

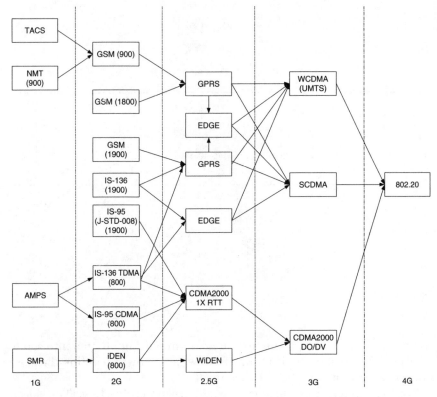

Figure 7.1 Wireless technology migration path.

The 802.20-compliant system will also interface with 802.11 Wi-Fi Systems via a smart phone through mobility management.

7.3 Integration

The integration of 802.20 into an existing mobile wireless system can be accomplished in many perturbations. The specific decisions and methodologies revolve around

1. What the long- and short-term objectives are?

2. When the process needs to begin and over what period?

3. How will it be implemented?

Since there are numerous possibilities to the perturbations, only a few of the possible situations will be covered. The small sampling is intended

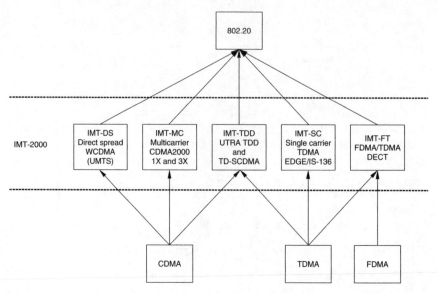

Figure 7.2   802.20 and IMT2000.

to help define and orient the management and design teams into some likely configurations or situations from which to further refine the specific requirements for the network itself.

The first integration consideration addressed the objective. (The objective factors in the obvious question of what do you want to do.) For instance, whether the offering of 802.20 integration will be for selected areas of the network, campus environments, or specific coverage areas for possible fleet management. Another consideration for defining the objective—what is the service or services desired to be offered—is a

TABLE 7.2   802.20 Technical Parameters

	Technical parameter
Wireless type	Mobile WAN
Frequency (Licensed spectrum)	400 MHz—3.5 GHz
Channel bandwidth	1.25 MHz, 5 MHz
Channel assignment	TDD, FDD
Mobility	250 km/hr—ITU-R M.1034-1
Min peak user data rate (Downlink (DL))	1 Mbps
Min peak user data rate (Uplink (UL))	300 kbps
Min peak aggregate data rate per cell (DL)	4 Mbps
Min peak aggregate data rate per cell (UL)	800 kbps
Security support	AES (Advanced Encryption Standard)
Compatible	IMT-2000, 802.11g

bold question to ask. Therefore, with the design process the fundamental issue of what you are trying to accomplish needs to be defined from the beginning.

The second main topic for defining the integration effort focuses on the timing of the process. The timing is important because it feeds the implementation decision or criteria. An example of timing could simply be that you want 802.20-compliant infrastructure in place for the core of the network within six months. Another alterative for timing could be that 802.20-complaint infrastructure needs to be operational in 80 percent of the existing cells sites within one year from the commencement of the project. The two differences in timing and fundamental objectives have a profound impact on the methodology used for implementing the system.

The implementation methodology from an infrastructure perspective involves the deployment method as well as the spectrum allocation scheme. For deployment the issue is what the specific deployment scheme needs to be to meet the marketing and business objectives, besides financial goals.

The three primary deployment methods involve

1. *1:1 deployment.* This requires that for every radio access site deployed, an 802.20-compliant site needs to be collocated at it. The tradeoffs involve time and cost as more infrastructure and capital, as well as operating costs increase.

2. *N:1 deployment.* This requires a more selective deployment of the new system allowing for the system to be deployed where marketing/sales believes the revenue or benefit will be most beneficial to the company. The tradeoff involves reduced cost with possible increased network complexity.

3. *Island deployment.* This deployment scheme involves deploying the service in very unique locations where they are effectively isolated networks (from each other). The advantage is that this is a very selective deployment, while the disadvantage lies in the actual integration constraints or issues with the rest of the mobile network.

Which deployment scheme will be used needs to be addressed at the start of the design cycle with marketing/sales and engineering debating the various combinations for advantages and disadvantages.

The other issue to consider addresses the spectrum allocation. Since the 802.20 system can have both FDD and TDD channel plans there is the possibility of having multiple channel bandwidth schemes deployed in a market or license area depending on the specific service-offering desires. The typical channel bandwidths are in increments of 1.25 MHz following CDMA2000 and Synchronous CDMA (SCMDA) channel allocations of FDD or TDD. Alternatively, the 802.20 system can be deployed as a 5 MHz aggregated channel or several 1.25 MHz channels comprising the 5 MHz channel bandwidth allocated for UMTS systems.

Figure 7.3 shows a possible spectrum segmentation scheme where 802.20 is deployed within an existing or planned CDMA2000 Network. The 802.20 spectrum has the same bandwidth as that used for CDMA2000 and depending on whether the deployment involves TDD or FDD, the possibilities only increase for possible configuration options.

However, for this example Fig. 7.3 shows a CDMA2000 system using 5 MHz of the spectrum, which is of course FDD. It is decided to deploy an 802.20 system for mobile data applications and the spectrum is divided to accommodate the new system requirements. For this situation it is assumed that the 802.20 system is also FDD but the spectrum cleared could involve an 802.20 TDD deployment offering unique services for each TDD channel.

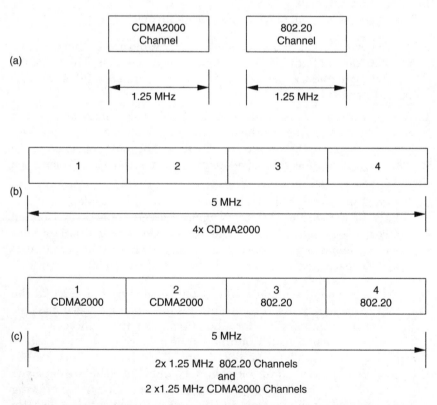

Figure 7.3 CDMA spectrum segmentation (*a*) CDMA2000 and 802.20 spectrum (*b*) CDMA2000 spectrum assignment, (*c*) Integration.

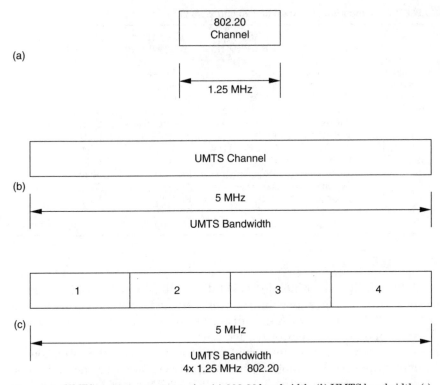

Figure 7.4 UMTS spectrum segmentation (*a*) 802.20 bandwidth, (*b*) UMTS bandwidth, (*c*) 802.20 integration into a UMTS channel.

For the example shown in Fig. 7.3 two 802.20 FDD channels are deployed in place of two CDMA2000 channels. Guard bands, if required, were not included in the example.

What follows next is an example of an 802.20 deployment within a UMTS, WCDMA, system in Fig. 7.4. For the UMTS system it is assumed that the 5 MHz of spectrum available is part of the overall spectrum that the operator has, otherwise it would be a complete system replacement and not an integration.

Figure 7.4 shows that the UMTS channel, assuming again FDD, can have four 802.20 FDD channels deployed within the same spectrum. Alternatively, there could be eight 802.20 TDD channels deployed in the spectrum allocated for a UMTS full duplexed channel.

7.4 Differences between 802.20 and 3G

What is the difference between 802.20 and 3G that falls under IMT-2000? The fundamental difference between 802.20 and IMT2000 is that

TABLE 7.3 Comparison between 802.20 and 2.5/3 G

	802.20	GSM-GPRS/EDGE	CDMA2000 1xRTT/EV-D0	UMTS
Frequency	<3.5 GHz	<2 GHz	<3.5 GHz	<3.5 GHz
Spectrum	Licensed	Licensed	Licensed	Licensed
LOS/NLOS	NLOS	NLOS	NLOS	NLOS
BW	1.25 MHz	200 kHz	1.25 MHz	5 MHz
	Packet	Circuit/packet	Circuit/packet	Circuit/packet
Channel plan	TDD/FDD	FDD	FDD	FDD
Modulation	OFDM	GMSK—TDM	CDMA	CDMA
Latency	Low	High	High	High
Type	Mobility	Mobility	Mobility	Mobility
	Symmetrical	Asymmetrical	Asymmetrical	Asymmetrical
Circuit switch/IP	IP	Both	Both	Both
Data	>1 Mbps	128/384 kbps	144 kbps/2 Mbps	144 kbps/384 kbps

802.20 is designed for mobile packet data and does not have the legacy issues with directly supporting circuit-switched services. For instance, CDMA2000 and WCDMA are all packet oriented systems that draw heavily upon their circuit-switched past.

802.20, however, was developed from the start with the intention to providing packet data mobility in a Wide Area Network. The system is designed to integrate into an existing 2.5G/3G Network by capitalizing on the packet core network that the wireless operator has or will have in place.

Table 7.3 attempts to quantify the main differences between 802.20 and several of the 2.5G/3G wireless access systems that are dominant.

Overall, 802.20 provides mobile packet data that integrate into IMT-2000 and provide a true universal wireless access technology that unifies all the 3G mobility platforms for true interoperability.

7.5   802.20 Compared to 802.16 and 802.11

Logic begs the question, "What is the difference between 802.20 and 802.16e and 802.11b/g?"

802.11, Wi-Fi, as discussed previously in the book is WLAN and is designed for fixed applications and not mobility. 802.11, whether it is a, b, g or any other variant, is designed for the LAN. 802.11 is an excellent standard leading to exceptional acceptance in the industry. However, 802.11x does not provide true mobile data, whereas 802.20 can.

However, unlike the 802.11 platform there is some overlap in functionality between 802.20, WBWA, and 802.16, WiMax. Both 802.20 and 802.16e are new radio access methods for wireless broadband. For instance,

TABLE 7.4   802.20, 802.11 and 802.16 Comparison

	802.11b	802.16a	802.16e	802.20
	LAN	MAN-PMP	MAN	MAN
Frequency	2 GHz	2-11 GHz	2-11 GHz	<3.5 GHz
Spectrum	Unlicensed	Licensed and unlicensed	Licensed	Licensed
LOS/NLOS	NLOS	NLOS	NLOS	NLOS
BW	20 MHz		5 MHz	1.25 MHz
	Packet	Packet	Packet	Packet
Channel plan	TDD	TDD/FDD		TDD/FDD
Latency	Low	Low	Low	Low
Type	Fixed	Fixed	Nomadic	Mobility
	Asymmetrical	Asymmetrical	Symmetrical	Symmetrical

802.16e is an extension or modification to 802.16a and is designed to provide IP mobility access in the 2 to 6 GHz licensed frequency bands. 802.20 also provides IP mobility but is destined toward the licensed frequency bands below 3.5 GHz, which are occupied by IMT-2000 compliant systems and 2.5G Networks.

Therefore 802.16e, like 802.20, is a *Wireless Metropolitan Area Network* (WMAN) access technology. But 802.16e is designed to integrate into an 802.16a network. 802.16e therefore is designed as an extension of the 802.16a network, like that designed for the *Multichannel Multipoint Distribution System* (MMDS) band that can support two-way mobility. It is specifically designed to provide a bridge between WLAN and WAN.

WBWA, 802.20 is designed to provide mobile data to a WAN that will be unique or integrated into an IMT-2000 series mobile network platform. In addition, the 802.20 mobile data system can also be deployed as a stand-alone system.

Table 7.4 shows a high-level difference between the various key 802 specifications. 802.15 is not included since it is meant for PAN applications.

References

http://grouper.ieee.org/groups/802/20

802.20-PD-02, "Mobile Broadband Wireless Access Systems: Approved PAR," IEEE, 2002.

802.20-PD-03, "Mobile Broadband Wireless Access Systems: Five Criteria (FINAL)," IEEE, 2002.

C802.20-03/48, "Channel Models and Performance Implications for OFDM-based MBWA," IEEE 802.20 Session no. 2, IEEE, 2003.

C802.20-03/50, "Overview of METRA Model for MBWA MIMO Channel," IEEE 802.20 Session no. 2, IEEE, 2003.

C802.20-03/ 61r1, "802.20 Technical Requirements—Coexistence Background," IEEE, 2003.

C802.20-03/63, "802.20 Evaluation Criteria—Ver 02," IEEE, 2003.

C802.20-03/69r1, "Draft 802.20 Permanent Document," IEEE, 2003.

C802.20-03/78, "802.20 Evaluation Criteria—Ver 05," IEEE, 2003.

Ansari, A., et al., C802.20-03/45r1, "Desired Characteristics of Mobile Broadband Wireless Access Air Interface," IEEE, 2003.

Goldhammer, M., 802.20—03/67, "Selected topics, including MAC+PHY aggregate capacity," IEEE, 2003.

Goldhammer, M.,C802.20-03/32, "Selected topics—Mobile System Requirements and Evaluation Criteria," IEEE, 2003.

Khan, F., C802.20-03/35, "Evaluation Methodology for 802.20 MBWA," IEEE, 2003.

Malenauer, J., C802.20-03/65r1, 16. IEEE Market Requirements for IEEE 802.20," IEEE, 2003.

Sohn, I., et.al., C802.20-03/49, "Comparison of SFBC and STBC for Transmit Diversity in OFDM System," IEEE 802.20 Session no. 2, IEEE, 2003.

Wilfson, J., et al., C802.20-03/47r1, "Terminology in the 802.20 PAR (Rev 1)," IEEE, 2003.

Xu, G., C802.20-03/46r1, "Channel Requirements For MBWA (Rev 1)," IEEE 8, 2003.

Youssefmir, M., et.al., C802.20-03/33r1, "Criteria for Network Capacity," IEEE, 2003.

Convergence Wireless Mobility and 802.x

The fundamental question that is driving the wireless industry is "Where is this all headed?" The answer in a simple word is *convergence*. However, this simple word has many different variations and meanings. Convergence at a high level is the intersection of data and circuit-switched services where the users no longer have to make the decision as to which transport method they need for the application. Carried further, it also is the merging of fixed and mobile phones, PDAs, and laptops so that the consumer has one device to use for multiple functions.

The choice of whether a single device or multiple devices should be used should be left to the consumer. However, the industry needs to provide one device that is configurable to the requirements of the customer. This is required because every customer has different needs. Also, these needs change over time and certain applications or features may not be needed on a regular basis. This single platform or chameleon platform is not far from reality.

The single platform is the envisioned X-Car where the subscribers purchase a *Software Definable Radio* (SDR) platform module and select the application modules they want to use. The subscriber, through a user friendly *graphical user interface* (GUI), instructs the SDR platform what services or functions are desired. This can be augmented through a personal module key that contains all the requisite authentication material.

Currently the fixed and wireless telecommunications industry is vying for the same market space where data and access to the Internet and or Intranet are provided by a single source, which also provides telephony services.

The convergence of 802.11 with 2.5G and 3G wireless mobile RAN is one of the first steps in the overall data and voice services convergence path.

8.1 Wi-Fi Integration

At the heart of Wi-Fi's integration with 2.5G/3G Networks is the fundamental proliferation of 802.11 that has seen widespread acceptance. Wi-Fi equipment for the end device as well as the infrastructure, APs, is relatively inexpensive, unlike its 2.5G/3G counterparts. Wi-Fi being inexpensive tends to imply that the service is easy to deploy and affordable. However, as always, the real issue regarding ease of deployment and affordability is that it depends on what the objective or service offering really is.

For instance, if Wi-Fi is deployed for a home network for linking a computer or computers to the Internet, then the low cost and ease of installation is real. But if the Wi-Fi Network is to support more than the home usage and provide ubiquitous access to public and private packet data networks, i.e., the Internet and corporate LANs, then the cost and installation concerns take on another level of complexity. Add in the component of wireless mobile services and the decisions that need to be made can be, and are, daunting. They are daunting because not all services are the same and you need to have prior knowledge of what you want to do before selecting the access method.

Some of the key issues that need to be addressed or factored into the decision process are:

- Coverage
- Roaming
- Backhaul
- Security
- Device

The coverage question comes down to the simple question, where are you going to use the service and how is it going to be used? If it is desired to have medical data available to visiting RNs for a large metropolitan area, the use of Wi-Fi by itself will need to be augmented by a 2.5G/3G system. However, if the objective is to have coverage in the visiting conference area then a stand-alone Wi-Fi System could be used.

The roaming issue that needs to be answered is akin to the issue of coverage. Is it desired to have the workforce use different Wi-Fi Networks that may or may not be integrated into a 2.5G/3G Network?

The issue of backhaul is important from a wireless service provider's as well as the end user's viewpoint. The service provider needs to determine and support the requisite backhaul transport level to ensure that the service offered, at the defined data rate, is achievable. This can be achieved via 802.16 RAN or traditional wired services. The operator is also compelled to try and offload heavy data traffic onto another RAN to preserve or better utilize the 2.5/3G RAN for less bandwidth-intensive services.

The end user needs to be assured that the services they plan on using can be accessed and used properly. If streaming video is required by users and they are mobile over a large geographic area, the implementation of this service may prove to be difficult. But on the other hand if the service fleet needs periodic updates, which are not necessarily in real time, then wide area may be the proper solution.

Though 2.5G and 3G Networks have many security levels built into them to protect against fraudulent usage, problems still occur. The Wi-Fi Networks, however, require that the end users also exercise some level of control over the security of their data and possibly their network.

The plethora of mobile data devices is actually a detractor for mobile data acceptance. All the current devices are excellent in one form or another. However, in most cases, to provide mobile service the RAN becomes secondary to the data entry method and the physical display, both of which fall under the form factor of the device.

There are multiple ways to link a 2.5G/3G Network to one or more Wi-Fi Networks in order to provide seamless broadband service. The method chosen depends on the 2.5G/3G RAN and the types of customers being pursued.

3G/Wi-Fi interoperability has been a focus of the wireless industry and is evident through the use of Mobile IP. Mobile IP, as discussed earlier, facilitates the continuity of the data session when the user is mobile, by facilitating handoffs between disparate networks so that they're not noticeable to users or programs, i.e., e-mail, when their IP connection changes.

There are two general classifications of mobile data users—casual and business. It is important to note that data requirements are very specific to the individual and cannot be a one-service-fits-all. Data requirements are unique and are vertical. For instance, your specific data requirements are different from my data requirements. This is evident from the type of device that is selected to be used, the applications selected, and the bandwidth required.

The ability to link a 2.5G/3G Network with Wi-Fi Networks is driven at this time by the network and the user device(s). Linking 2.5G/3G Networks with Wi-Fi Networks involves both business and technical considerations. The business considerations focus on the Service Level

Agreements (SLA) and possible charging rates that can or will be used. This is in addition to the determination of what services specifically will be used or available.

The technical considerations involve design, deployment, provisioning, and ongoing support. Technical issues generally involve billing and authentication, and increase in complexity when the networks are owned by different companies. Therefore the demarcation and how the networks will interact with each other in the delivery of services is another area of technical interaction. But the ongoing support is of course a critical decision that involves problem identification, resolution, and enhancements.

The wireless data provider must therefore be able to authenticate the user, track the usage, and then deliver that information in a format compatible with the operator's billing systems.

Mobile wireless providers are trying to reduce churn by creating sticky content to keep their customers. One such method is the use of an integrated bill besides data services and the possibility of having a single monthly bill for 3G and Wi-Fi is a convenience that could be one more differentiator to reduce churn.

8.2 Commonality between WCDMA/ CDMA2000

Both WCDMA and CDMA2000 have several commonalities which are part of the IMT2000 platform specification. Both systems use CDMA technology with CDMA 2000 requiring 2.5 MHz while WCDMA needs 10 MHz of spectrum. Both systems will be able to interoperate with each other and it is possible for a wireless operator to deploy both a CDMA2000 Network as well as a WCDMA system barring of course the capital cost issues.

Both systems have a migration path from existing 2G platforms to that of 3G. However, the path both systems take is different. This is driven by the embedded infrastructure the operator has already deployed. Since the endgame is to offer high-speed packet data services to the end user, the real issue between both of these standards within the IMT2000 specification is the methodology for realizing the desired speed.

WCDMA uses a wideband channel while CDMA2000 uses both a wideband and several narrowband channels in the process of achieving the required throughput levels.

Additionally, both WCDMA and CDMA2000 are designed to operate in multiple frequency bands. Both systems can operate in the same frequency bands provided the spectrum is available.

Therefore, the commonalities between WCDMA and CDMA2000 can be summed up in the brief bullet points that were introduced at the beginning of this chapter.

- Global standard
- Compatibility of service within IMT-2000 and other fixed networks
- High quality
- Worldwide common frequency band
- Small terminals for worldwide use
- Worldwide roaming capability
- Multimedia Application Services and terminals
- Improved spectrum efficiency
- Flexibility for evolution to the next generation of wireless systems
- High-speed packet data rates
  - 2 Mbps for fixed environment
  - 384 Mbps for pedestrian
  - 144 kbps for vehicular traffic

8.3 Software Defined Radio (SDR)

Software definable radios have been available for many years but have seen most of their use focused on military applications. The commercial world has long sought an SDR but the cost, until recently, has been the inhibiting factor for its potential use.

An SDR would enable a mobile phone to support a vast array of RAN. In fact, an SDR could be used to provide one device to the subscriber that would interface with any of the multitude of RAN protocols such as UMTS, GSM, CDMA2000, AMPS, and DAMPS.

The SDR would be able to support multimodes, modulation schemes, as well as support the interface to many services. For instance, an SDR could be used by a wireless operator who has several RAN technologies in diverse markets facilitating roaming from one market to the other without concern about the underlying RAN in each market.

An SDR is capable of covering the wide range of spectrum used for mobility—400 MHz to 2.4 GHz and beyond. The SDR will also be able to support 802.11b/g access methods, further enhancing the mobile data environment.

The use of an SDR enables the use of a single platform for all mobile devices. The SDR would enable subscribers to purchase one device that allows them to configure it to meet their specific user requirements. This method would enable subscribers to tailor their specific data and mobile requirements to match their specific needs instead of making numerous compromises.

However, the SDR, due to its wide receiver bandwidth, will be more susceptible to interference and possible range reductions. But the ubiquity and overall functionality of the device has the exceptional potential of enhancing the customer's overall experience with fixed and mobile data.

8.4 Ultra Wideband (UWB)

Ultra wideband also referred to as UWB is an emerging wireless access method. UWB falls under the 802 suite of standards namely 802.15 and will operate in the unlicensed bands. UWB is a *Personal Area Networks* (PAN) technology that is intended to improve the throughput capability now offered through use of Bluetooth. Presently, Bluetooth can support PAN data rates of about 1 Mbps but UWB has the possibility of offering rates exceeding 480 Mbps.

There are some unique advantages with UWB due to its unique propagation properties. More specifically, due to the short time duration of the signal it can support extremely high data rates supporting multiple users in a small area, PAN. The short duration of the signal and the resulting modulation and demodulation scheme makes it very immune, at short range, to multipath interference that is a predominant problem in both mobile as well as in building situations.

The specific access method that will be used with UWB is undefined at this moment with the *Institute of Electrical and Electronic Engineers* (IEEE) and *International Telecommunications Union* (ITU) apparently supporting two different standards for the new and improved PAN. But the potential for UWB is vast and the range of applications is still being put together.

8.5 Business Considerations

At the heart of the convergence decision lie the business considerations which directly influence the deployment and implementation decisions of a company. For example, what are the services that will be or can be offered?—a simple question but often it is extremely difficult to answer. The services selected should match the market that is being sought, which is obvious. However, how will the services selected be distributed, i.e., selective or global? Will the services offered only be selectively targeted to an island of interest, niche, or will the service be advertised for the entire market? There are technical and financial issues with both of these approaches.

The primary objective of any wireless company, or any company, is not to explore technology but to make money. The governing method for determining the money made is driven by the following equation.

$$\text{Gross Profit} = \text{Gross Revenue} - \text{Gross Expenses}$$

and

$$\text{Gross Profit} > 0$$

While this may seem primitive in nature, the fundamental issue is that this concept is not fully understood. As it has been witnessed many times in the wireless industry, technology platforms and systems have been converted based on the quest for technology instead of answering the three questions.

- Who is the customer?
- What do they really want?
- What will they be willing to pay for this service?

Failure to answer these questions can and often does lead to the deployment of technology for technology's sake. More specifically, if you have an APRU (average revenue per user) = $50 US but the service costs $51 to deliver including capital expenditures, then its viability is questionable.

Ultimately the decision comes down to vision—what is your vision for the future? You must have a firm understanding of where the company is today and what it will become when it grows up. Again another profound question but one that needs to be resolved at the beginning of any major shift in network architecture or direction. The marketing department should be at the center for helping provide the vision. The vision for the company and how it envisions where the revenue will come from is the billion dollar question that has to be answered.

The four basic steps are the same as those that govern marketing and they are referred to as the four Ps—product, place, promotion, and of course price.

- *Product.* This involves defining the services offered, installation, warranty, product lines, packaging, and branding, to mention the most important.
- *Place.* This refers to the marketing objectives, i.e., penetration, the channels used for achieving the penetration (direct and indirect); market exposure and competition; distribution methods (direct and indirect); *value added reseller* (VAR) partnerships; distribution of add-on services and products; and sales locations like direct and indirect stores.

- *Promotion.* This basically involves the bundling or unbundling method, direct and indirect sales forces (number, training, incentives), advertising (media type, copy trust, trade shows, press, ad types, agency—if used). Also included is the sales promotion involving commissions, customer retention programs, discounts, VAR products.

- *Price.* This involves the objective which can be to ensure profitability for every product offering or have a loss leader in the mix. Flexibility in terms of pricing for volume purchases or bundling techniques, product life cycle pricing, local market pricing differentiation, and sales allowances for indirect sales or CPE replacement and or subsidization.

All of the four Ps center around the important issue of enticing the customer to want the service perceive value with the product, use the service, and continue to use the service and possibly utilize additional services either offered now or some time in the future.

From the inception, the marketing department needs to feed many groups within the company. The sales department obviously needs to know the vision and product offerings that will be offered for initial sale. The technical community needs information related to traffic or capacity planning to ensure that the factory is in place for the sales force to sell while minimizing the capital and operating expenses. The customer care, billing, and provisioning groups need to know the market and services so that bills can be issued and of course the customer serviced in as high a quality as economically possible, based on the services offered.

Marketing information is not static but is dynamic in that market conditions always change and the company must respond to those changes, the response could be not to act. Therefore, the marketing department needs to interact with the other functioning entities within the company to ensure that continued success and refinements take place.

From the aspect of technical communities the marketing information will drive the infrastructure layout and design. The RF and network design aspects while related are not the same but use the same fundamental information to derive their answers and designs.

The overriding point is that convergence is not only a technology issue but also an approach that the wireless companies need to incorporate in their internal workings as well.

8.6 Services

Services, services, services. The objective of convergence is to provide more services that will be purchased and used by customers. Technology is nice but if the technology deployed does not fulfill a need in the market then it will, no matter how good it is, fail either to be profitable or be replaced with something else that meets the demand.

The onslaught of 2.5G/3G and Wi-Fi capabilities introduce a multitude of possibilities, all centered on the IP. The use of IP has been predominantly centered on fixed locations. Wi-Fi allows for some level of mobility but is usually contained to a campus or home network environment. The integration of wireless mobility with 802.11 enables the possibility of numerous service offerings.

The services offered are partly dependent on the RAN but more on the device itself. This is a fundamental shift in service offering from the past in that the RAN was the driving force in mobility for service. But with the proliferation of IP and the ever maturing RAN for mobility the next real quest for service offering and enhancements is device driven.

In order for the mobile devices to achieve real integration with Wi-Fi, and possibly WiMax, it will still be necessary to achieve the following with regards to the handset or user device.

1. More stable software for the devices, no *Blue Screen of Death* (BSOD)

2. Simplify choices and eliminate the configuration headaches

3. Truly eliminate the interoperability issues between networks

4. Improve data security by protecting the information in the device as well as the network

5. Improve the technical support skill for data integration

6. Battery life improvements—as more is demanded from a device the battery life diminishes

In addition to the mobile device issues for enhancement of the user's experience it will be necessary to inform the user when a call drops or a session is terminated prematurely during a download. The user needs to be informed that the transaction was completed or not completed.

With the above said, the desire of any operator is to offer and support real applications which provide some sticky content that helps keep the customer. The specific services that can and will be offered by any mobile wireless operator are of course market and customer dependent.

However, the following are some of the possible services that should be considered in the race for convergence between 2.5G/3G and Wi-Fi.

8.6.1 Home networking

This is where the hybrid phone and access method can be pursued. In this situation the 2.5G/3G phone acts as a mobile or cordless phone depending on the location where the call is placed.

The mobile phone can also link to a laptop via Bluetooth and enable both voice and data services to be used at the same time. The rating method or methodology used can be based on position location information. When subscribers are close to their residence or place of business within a defined range the access is considered to be a local/landline connection, otherwise it is considered to be a typical mobile. Due to interference potential the use of Bluetooth in conjunction with 802.11 needs to be made carefully.

8.6.2   802.11 and VoIP

The use of VoIP can be extended with the integration of 2.5G/3G within a corporate environment. The service offering would be centered on the ability of the mobile device to communicate for all voice-related functions via the packet network for all voice calls. The packet call would be routed to the *Internet Protocol Private Branch Exchange* (IPPBX) for the corporation when the employee is away from the office.

When the employee was in the office the phone would sense the 802.11 AP and use it, instead of the 2.5G/3G Network for transporting the IP call.

8.6.3   Hot spot (802.11/802.16)

The use of wireless mobility in conjunction with 802.11 hot spots will be essential. The ability to access a network that is not part of the home 802.11 Network will be necessary to achieve true ubiquity.

This is also important from an operator's perspective to preserve the radio resources of the 2.5G/3G Network. When a subscriber begins a data-intensive session the mobile device should look for an 802.11 or 802.16a AP for data connectivity.

8.6.4   Location based

There are an infinite amount of location-based services that could be used and integrated with 2.5H/3G and Wi-Fi. One possibility is that location-based services can be tied into Wi-Fi hot spots where streaming video clips are shown to a mobile user when close to an AP.

8.6.5   Interactive games

The increased prevalence of interactive games that are used by mobile phone users can be capitalized on. With Bluetooth connectivity a mobile user could interactively use PS2, X-Box or Game Cube while on a long trip in a vehicle or waiting at an airport. The interactivity and performance could be enhanced through Mobile IP and Wide Area 802.11 Systems.

8.6.6 Variable prepay

The need to tailor data services to each individual client or corporation can be enhanced through the use of prepay. More specifically, the customer could use the standard voice offering and then prepay for data usage, thereby limiting initial exposure to this service.

8.7 Budgeting

The budgeting aspects to the plan need to account for all capital and operating expense items associated with the wireless mobility and access point elements. The budget aspects for the plan need to identify the total cash requirements for the planning period. Therefore, for any successful and meaningful plan the identification of the capital and operating expenses needs to be defined.

The actual format for the budgeting issues associated with the plan should correlate with the format used for the technical budget submitted for the time frame covered. The budgeting aspects need to include all new capital and expense requirements. The new capital and expense requirements need to be compared against the current budget already submitted and variances identified.

The capital and operating expenses for the budget process will typically be asked for on a monthly basis. The operating expenses can and should be broken down into monthly increments. However, the capital spending is much harder to define on a monthly basis. Therefore, it is best to list the capital requirements on a quarterly basis which in reality is the best approach.

The major components of the capital budget are identified next. It is important to note that the list of major capital items include the costs of the *Fixed Network Equipment* (FNE), hardware and initial software load, and installation, as well as all associated construction costs.

1. Cell site
2. Tower, TI
3. Switching Centers (MSC)
4. Data Network Connectivity (private/public)
5. Packet Data Platforms (PDSN/SGSN)
6. Backhaul Microwave Systems (802.16)
7. Ancillary Switching and Packet platforms
8. NOC

Operating expenses need to include all the expenses to keep the system running, including the ability to deliver the various voice and data services. They are usually associated with the backhaul or interconnection leased line costs.

The operating expenses also include software maintenance agreements, even though the software loads may include bug fixes. Other important, but not all inclusive, topics for the operating expenses include ac/dc power, cell and core network maintenance, landscaping, and HVAC maintenance, to mention a few.

Typically, the finance department has a set of guidelines and codes that they use in presenting the capital and expense budgets. The inclusion of the finance department format ensures understanding between the two departments.

When putting together the budget for the AP or service offering it is important that realistic values and expectations are used. The desire to show everything being built as a profitable endeavor is noble but not necessarily accurate. There have been many instances where a system was deployed knowing full well that it would not be individually profitable; however, it enhanced the overall network service offering and kept the operator competitive. Of course the true desire and objective is to install and maintain the operating system so that it is and remains profitable. In addition, there are some difficult-to-identify savings associated with the implementation of a wireless LAN. These include increased productivity, better employee interaction, improved communications with customers, and faster decision making.

In order to ensure the overall success of a company, plans are required. For any technical deployment or project it is essential to put forth a plan that defines what you are planning and doing, why, when, and how. To properly communicate the plan it is essential to follow a guideline.

The RAN Plan Report is the part of the project where all the efforts put forth to date are combined into a uniform document. There are many methods and formats that can be used for putting together a report. When crafting the final report it is extremely important to remember your target audience. The report itself will be used by both upper management and the engineering department to conduct the actual planned network deployment.

The output of the design effort will be a report that will need to follow a uniform structure that captures the salient issues as well as facilitates the design review process. The outline shown is a recommended format to follow. However, your particular system requirements will, in all likelihood, require some variations to the proposed structure.

1. Executive summary

2. Introduction

3. Subscriber forecast

4. Expansion or migration plan

5. RF System

6. Circuit switch

7. PDSN

8. Interconnection

9. IP scheme

10. Implementation plan

11. Headcount requirements

12. Budget

13. ROI

Regardless of the actual format used the process is important to follow and, as said numerous times, essential for success.

8.8 Convergence

The mobile wireless community is upgrading and deploying a host of wireless platforms that are focused on lowering the overall cost of ownership for the operator as well as providing new and improved packet services.

While currently the proliferation of 802.11b/g is beginning to influence the customer's overall data experience by untethering them it has seen widespread acceptance in both the commercial as well as residential markets. 802.16 and 802.16a will also factor in the reduction of the total cost of ownership. It is unclear at this moment what impact 802.20, UWB, and 802.15 for mobile data, will have on convergence.

The IMT2000 overall objective is to provide high-speed data packet services, offered to the customer through a Radio Access Network. The RAN can either be a stand-alone network or use several other RAN standards to augment the service offering. The augmented service offering is designed to improve the overall customer experience.

Mobile phones are now becoming so commonplace that it is the primary communication device for a customer or family, with traditional landline services either not being used or simply providing backup.

The convergence of 802.11 and wireless mobility has several possible directions that it can take. Figure 8.1 shows the possibility of 802.20 as the overall convergence point for wireless mobility. But 802.20 is still in the development stage at this time. There are, however, other alternatives for convergence.

Figure 8.2 is an example of how 802.11 can and is converged with an existing GSM Network that uses *General Packet Radio Service* (GPRS)

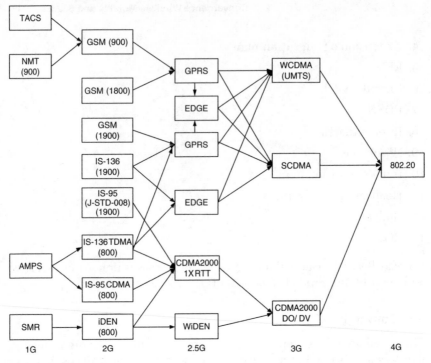

Figure 8.1 4G.

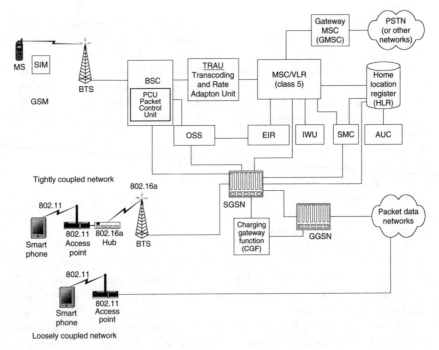

Figure 8.2 GSM and 802.11 convergence.

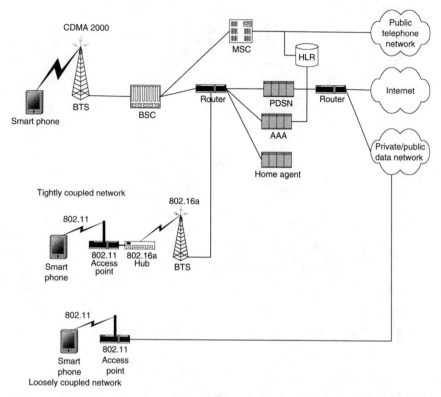

Figure 8.3   CDMA2000 and 802.11 convergence.

and or *Enhanced Data Rates for Global Evolution* (EDGE). The 802.11 systems can be integrated directly into the operator's network or through a roaming agreement provide access to the GSM subscriber.

Figure 8.3 is similar in concept to Fig. 8.2 with the exception that the underlying RAN is CDMA2000.

Figure 8.4 is an example of how a wireless operator that migrates from GSM to UMTS integrates with an 802.11 system. UMTS's connectivity and integration with 802.11 systems is relatively the same as that for GSM due to the commonality of the core packet network.

Figure 8.5 is rather unique in that it highlights the possibility of convergence of several wireless mobility RANs with 802.11. This configuration can be exploited to support international roaming or acquisitions that involve different RANs. There are numerous commonalities that can be exploited in the configuration shown in Fig. 8.5, like the sharing of a common HLR.

Finally, Fig. 8.6 is an example of how soft switches can play a positive role in the convergence of mobile and fixed wireless data.

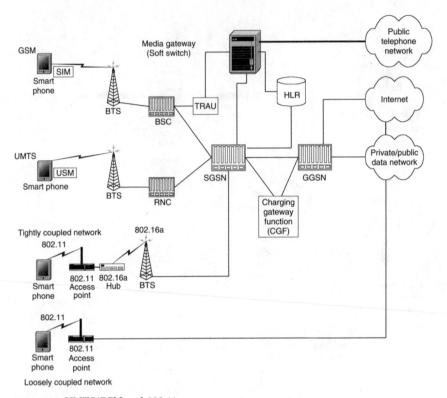

Figure 8.4   UMTS/GSM and 802.11 convergence.

8.9   Benefits of Convergence of Wi-Fi and Wireless Mobile

To some extent Wi-Fi can be thought of as a competitor to mobile services, in particular, the 3G high-speed data access capabilities. On the other hand, the services are very complimentary. While Wi-Fi provides higher data rates mobility provides much greater roaming and voice services on a global basis. There are several benefits to the customer as well as the mobile operator when the two technologies are combined.

The customer, for example, can be a corporation if they desire to replace much, if not all, of their premise telephone system with IP-based Wi-Fi-enabled handsets. If then these handsets are also capable of mobile service they have a single solution for both landline and mobile phone service for the workforce. This provides high-speed access on less expensive Wi-Fi equipment while in the building and mobility and 3G data services on the road.

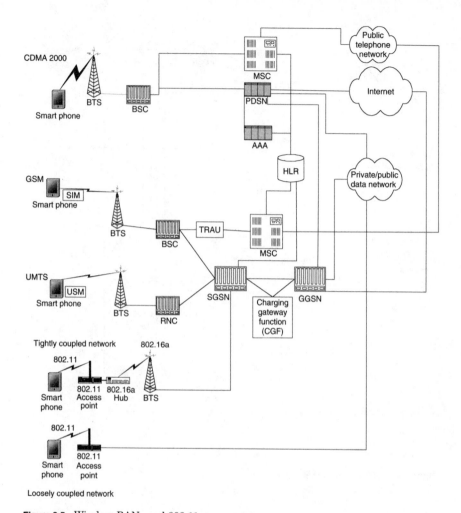

Figure 8.5 Wireless RANs and 802.11 convergence.

The wireless operator will benefit from churn reduction. If customers bundle their mobile and local phone solutions, they are far less likely to change service providers. Another benefit is the reduction in use of the operator's spectrum. Spectrum is of course a very scarce resource.

There are a number of unknowns as to how this would work commercially.

Will a customer pay for Wi-Fi bandwidth? Reluctantly at best.

How will a mobile operator offer local service as well? Through *Regional Bell Operating Companies* (RBOCS) like *Southern Western Bell Communications* (SBC)?

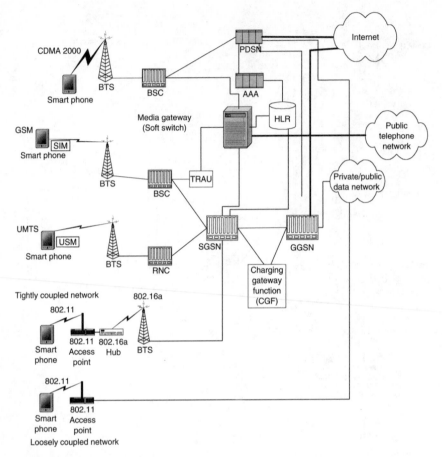

Figure 8.6 Soft switch convergence.

When will the 802.11/Mobile Wireless handset or other device be prolific?

How does this all fit in with the once again predicted shift to VoIP?

In closing, as you can see and imagine there are numerous possibilities related to the methods and possibilities for convergence to take place with 802.11 and wireless mobility. The key concepts that need to be kept in mind with convergence are:

- What are you really trying to accomplish?
- What is the optimal technology?
- When is the solution required?
- How will you achieve the deployment plan?

References

Dhir, A., "The ABC's of 2.4 and 5 GHz Wireless LANS," Xilinx, August 1, 2001.

Gary, J., "Welcome to the Jungle," *Electronics Design News*, October 30, 2003, p. 39.

Kent III, S. D., "Crafting Software-Defined Radio for 3G," *Electronic Engineering Times*, November 17, 2003, p. 77.

Mannion, P., "UWB Fate Still Up in the Air," *Electronic Engineering Times*, December 8, 2003, pp. 18–26.

Mannion, P., "Wireless Mesh Networks Gain Traction," *Electronic Engineering Times*, November 17, 2003, p. 56.

Marek, S., "Fragmented Wi-Fi Industry Begins to Coalesce," *Wireless Week*, October 15, 2003, pp. 7–8.

Mayor, T., "Wi-Fi Hot Spots Head Up," *Electron Business*, May 1, 2003, pp. 14–16.

O'Shea, D., "A marriage of conveience: Where Wi-Fi & Mobile merge," *Telephony*, March 17, 2003, pp. 24–36.

Smith, B., "3G What Lies Ahead," *Wireless Week*, March 15, 2003, pp. 6–8.

Walters, M., "GPRS Elixir," *Wireless Review*, January 2002, p. 48.

Index